ASPECTS OF CHEMICAL EVOLUTION

XVIIth SOLVAY CONFERENCE ON CHEMISTRY

ADVANCES IN CHEMICAL PHYSICS
VOLUME LV

ASPECTS OF CHEMICAL EVOLUTION

XVIIth SOLVAY CONFERENCE ON CHEMISTRY
WASHINGTON, D.C., APRIL 23–APRIL 24, 1980

Edited by

G. NICOLIS

University of Brussels · Brussels, Belgium

ADVANCES IN CHEMICAL PHYSICS
VOLUME LV

Series editors

Ilya Prigogine
University of Brussels
Brussels, Belgium
and
University of Texas
Austin, Texas

Stuart A. Rice
Department of Chemistry
and
The James Franck Institute
University of Chicago
Chicago, Illinois

AN INTERSCIENCE® PUBLICATION

JOHN WILEY & SONS
New York · Chichester · Brisbane · Toronto · Singapore

An Interscience® Publication

Copyright © 1984 by John Wiley & Sons, Inc.

Library of Congress Cataloging in Publication Data:

Solvay Conference on Chemistry (17th : 1980 : Wash-
    ington, D.C.)
    Aspects of chemical evolution : XVIIth Solvay
Conference on Chemistry, Washington, D.C., April 23–
April 24, 1980.

    (Advances in chemical physics ; v. 55)
    "An Interscience publication."
    Includes index.
    1. Chemical evolution—Congresses. 2. Instituts
Solvay—History—Congresses. I. Nicolis, G.,
1939–   . II. Title. III. Series.

QD453.A27 vol. 55 [QH325]   541s   [577]   83-12386
ISBN 0-471-88405-7

Printed in the United States of America

10 9 8 7 6 5 4 3 2 1

INTRODUCTION

Few of us can any longer keep up with the flood of scientific literature, even in specialized subfields. Any attempt to do more, and be broadly educated with respect to a large domain of science, has the appearance of tilting at windmills. Yet the synthesis of ideas drawn from different subjects into new, powerful, general concepts is as valuable as ever, and the desire to remain educated persists in all scientists. This series, *Advances in Chemical Physics,* is devoted to helping the reader obtain general information about a wide variety of topics in chemical physics, which field we interpret very broadly. Our intent is to have experts present comprehensive analyses of subjects of interest and to encourage the expression of individual points of view. We hope that this approach to the presentation of an overview of a subject will both stimulate new research and serve as a personalized learning text for beginners in a field.

ILYA PRIGOGINE

STUART A. RICE

PREFACE

The reports and discussions contained in this volume have been presented at the XVIIth Solvay Conference on Chemistry. This was a special Solvay Conference organized jointly by the Solvay Institutes and the National Academy of Sciences of the United States, on the occasion of the 150th anniversary of the establishment of Belgium as an independent state. It was held in Washington, D.C., on April 23–24, 1980.

The first part of the Conference was devoted to the historical role of the Solvay Institutes, and in particular of the Solvay Conferences on Physics and Chemistry, in promoting international scientific cooperation. The second part was devoted to some aspects of chemical evolution. This subject comprises an impressive number of different, yet complementary, questions having an interest in their own right.

At the molecular level, it is believed that atmospheric conditions prevailing on earth some 4 billion years ago triggered an evolution of simple molecules to more complex forms. Eventually this led to biopolymer formation, which is at the basis of the origin of life. Somewhere along this path transition points such as chirality and self-replication arose, which conferred on evolution a markedly irreversible character.

At the macroscopic level, evolution brought a drastic change in the environment. For instance, the earth's atmosphere after the appearance of life is completely different than before. In addition, at this level, the analysis of physical, chemical, and biological problems shows that the evolution to complexity is accompanied by the successive breaking of symmetries of various kinds induced by two elements that continuously recur throughout the different stages: *nonequilibrium constraints*, necessary to enable the system to pump the appropriate raw materials from its environment; and *nonlinearities* in the form of feedbacks, necessary to speed up what would be otherwise an exceedingly slow rate of evolution. Finally, at the crucial moment of choice of a particular pathway among the multitude available, *fluctuations* play a crucial role.

We hope that by bringing together presentations referring to all these different facets of evolution, the present volume will contribute to the development of an interdisciplinary approach to this fascinating field.

The late Philip Handler, president of the U.S. Academy of Sciences, played a key role in the success of the Conference. Thanks are also due to Professor Bryce Crawford, Home Secretary of the Academy, and his

vii

staff for the efficient handling of the various problems posed by the organization of the meeting.

ILYA PRIGOGINE

GREGOIRE NICOLIS

Brussels, Belgium
July 1983

ix

W. LIPSCOMB, Harvard University, Department of Chemistry, Cambridge, Massachusetts, U.S.A.

R. MARTIN, Université Libre de Bruxelles, Service de Chimie Organique, Bruxelles, Belgium

J. E. MAYER, University of California, San Diego, Department of Chemistry, Revelle College, La Jolla, California, U.S.A.

YU. OVCHINIKOV, Shemyakin Institute of Bioorganic Chemistry U.S.S.R. Academy of Sciences, Moscow, U.S.S.R.

L. PAULING, Linus Pauling Institute of Science and Medicine, Menlo Park, California, U.S.A.

V. PRELOG, Eidgenössische Laboratorium für Organische Chemie, Zürich, Switzerland

LORD TODD, Christ's College, Cambridge, England

G. WITTIG, Organisch-Chemisches Institüt der Universität, Heidelberg, Federal Republic of Germany

Participants

E. AMALDI, Istituto di Fisica "Guglielmo Marconi," Universita degli Studi, Piazzale delle Scienze, Rome, Italy

G. AUCHMUTY, Indiana University, Department of Mathematics, Bloomington, Indiana, U.S.A.

A. BABLOYANTZ, Université Libre de Bruxelles, Service de Chimie Physique II, Bruxelles, Belgium

J. BELEW, Vice-President, Baylor University, Waco, Texas, U.S.A.

A. BELLEMANS, Université Libre de Bruxelles, Campus Plaine, Bruxelles, Belgium

A. BERGER, Institut d'Astronomie et de Géophysique, Université Catholique de Louvain, Louvain-La-Neuve, Belgium

J. BIGELEISEN, Assembly Chairman, National Research Council, Assembly of Mathematical and Physical Sciences, State University of New York at Stony Brook, Stony Brook, New York, U.S.A.

F. BINGEN, Faculteit van Wetenschappen, Vrije Universiteit Brussel, Brussel, Belgium

R. M. BOCK, The Graduate School, University of Wisconsin, Madison, Wisconsin, U.S.A.

S. CLAESSON, Uppsala Universitet, Fysikalisk-Kemiska Institutionen, Uppsala, Sweden

M. DOLE, Baylor University, Department of Chemistry, Waco, Texas, U.S.A.

M. EIGEN, Max-Planck-Institut für Biophysikalische Chemie, Nikolausberg, Göttingen, Federal Republic of Germany

T. ERNEUX, Université Libre de Bruxelles, Service de Chimie Physique II,

Bruxelles, Belgium, and Northwestern University, Department of Applied Mathematics, Evanston, Illinois, U.S.A.

P. FIFE, University of Arizona, Department of Mathematics, Tucson, Arizona, U.S.A.

G. FONKEN, Vice-President for Research, The University of Texas, Austin, Texas, U.S.A.

H. FRISCH, State University of New York, Department of Chemistry, Albany, New York, U.S.A.

M. GHIL, New York University, Courant Institute of Mathematical Sciences, New York, New York, U.S.A.

P. GLANSDORFF, Université Libre de Bruxelles, Campus Plaine, Bruxelles, Belgium

A. GOLDBETER, Université Libre de Bruxelles, Service de Chemie Physique II, Bruxelles, Belgium

T. A. GRIFFY, The University of Texas, Department of Physics, Austin, Texas, U.S.A.

T. GUO, I.I.T., Research Institute, Annapolis, Maryland, U.S.A.

W. GUO, I.I.T., Research Institute, Annapolis, Maryland, U.S.A.

B. L. HAO, Chinese Academy of Sciences, Institute of Theoretical Physics, Beijing, People's Republic of China

H. HERS, Université Catholique de Louvain, Faculté de Médecine, Laboratoire de Chimie Physiologique, Bruxelles, Belgium

B. HESS, Max-Planck-Institut für Ernährungsphysiologie, Dortmund, Federal Republic of Germany

J. HIERNAUX, Université Libre de Bruxelles, Service de Chimie Physique II, Bruxelles, Belgium

H. D. HOLLAND, Harvard University, Cambridge, Massachusetts, U.S.A.

E. IMMERGUT, Scientific American, New York, New York, U.S.A.

K. KALTHOFF, The University of Texas at Austin, Department of Zoology, Austin, Texas, U.S.A.

L. D. KAPLAN, National Aeronautics and Space Administration, Goddard Space Flight Center, Greenbelt, Maryland, U.S.A.

E. KATCHALSKY, The Weizmann Institute of Science, Rehovot, Israel

S. KAUFFMAN, University of Pennsylvania, Department of Biology, Philadelphia, Pennsylvania, U.S.A.

D. KONDEPUDI, Center for Studies in Statistical Mechanics, The University of Texas, Austin, Texas

P. LASZLO, Institut de Chimie, Université de Liège, Liège, Belgium

R. LEFEVER, Université Libre de Bruxelles, Service de Chimie Physique II, Bruxelles, Belgium

E. MARGOLIASH, Northwestern University, Department of Biochemistry and Molecular Biology, Evanston, Illinois, U.S.A.

R. Martin, Université Libre de Bruxelles, Service de Chimie Organique, Bruxelles, Belgium

J. E. Mayer, University of California, San Diego, Department of Chemistry, Revelle College, La Jolla, California, U.S.A.

M. McElroy, Harvard University, The Center of Earth and Planetary Physics, Cambridge, Massachusetts, U.S.A.

J. Mehra, 7830 Candle Lane, Houston, Texas, U.S.A.

M. Migeotte, Institut d'Astrophysique, Université de Liège, Cointe-Ougrée, Belgium

S. Miller, Department of Chemistry, University of California, San Diego, La Jolla, California, U.S.A.

M. Nicolet, Université Libre de Bruxelles and Department of Electrical Engineering, Pennsylvania State University, University Park, Pennsylvania, U.S.A.

G. Nicolis, Université Libre de Bruxelles, Campus Plaine, Bruxelles, Belgium

G. North, Physics Department, University of Missouri, St. Louis, Missouri, U.S.A.

R. Noyes, Department of Chemistry, University of Oregon, Eugene, Oregon, U.S.A.

C. Ponnamperuma, Laboratory of Chemical Evolution, Department of Chemistry, University of Maryland, College Park, Maryland, U.S.A.

I. Prigogine, Université Libre de Bruxelles, Service de Chimie-Physique II, Bruxelles, Belgium, and University of Texas, Center for Studies in Statistical Mechanics, Austin, Texas, U.S.A.

R. A. Reck, Physics Department, General Motors Research Laboratories, Warren, Michigan, U.S.A.

S. A. Rice, James Franck Institute, University of Chicago, Chicago, Illinois, U.S.A.

J. Ross, Stanford University, Department of Chemistry, Stanford, California, U.S.A.

A. Sanfeld, Université Libre de Bruxelles Service de Chimie-Physique II, Bruxelles, Belgium

J. Schell, Max-Planck-Institut für Suchtungsforschung, Cologne, Federal Republic of Germany

E. Schoffeniels, Université de Liège, Faculté de Médecine, Laboratoire de Biochimie Générale et Comparée, Institut Léon Frédéricq, Liège, Belgium

P. Schuster, Institüt für Theoretische Chemie und Strahlenchemie der Universität Wien, Wien, Austria

S. Spiegelman, College of Physicians and Surgeons of Columbia University, Institute of Cancer Research, New York, New York, U.S.A.

J. Stucki, Pharmakologisches Institut, Berne, Switzerland

H. Swinney, University of Texas, Department of Physics, Austin, Texas, U.S.A.

C. Sybesma, Vrije Universiteit Brussel, Brussel, Belgium

V. Szebehely, University of Texas, Department of Aerospace Engineering and Engineering Mechanics, Austin, Texas, U.S.A.

R. Thomas, Université Libre de Bruxelles, Département de Biologie Moléculaire, Rhode Saint-Genèse, Belgium

J. S. Turner, University of Texas, Center for Studies in Statistical Mechanics, Austin, Texas, U.S.A.

A. R. Ubbelohde, Imperial College, Department of Chemical Engineering and Chemical Technology, London, England

G. Van Binst, Faculteit van Wetenschappen, Organische Chemie, Vrije Universiteit Brussel, Brussel, Belgium

R. Zander, Institut d'Astrophysique, Université de Liège, Cointe-Ougrée, Belgium

Participants from U.S. Governmental and Other Research Institutions

A. C. Aikin, Senior Scientist, Goddard Space Flight Center

W. R. Benson, Director, Division of Drug Chemistry, F.D.A.

N. Bikales, Program Director, Polymers, N.S.F.

D. Challinor, Assistant Secretary, Smithsonian

D. Dendrinos, Visiting Fellow, D.O.T.

J. D. D'Ianni, President, A.C.S.

R. P. Epple, D.O.E.

A. F. Findeis, Head, Chemical Synthesis and Analysis Section, N.S.F.

D. L. Gilbert, Physiologist, N.I.H.

M. L. Good, A.C.S. Board

D. M. Green, Program Director, Biochemistry, P.C.M.-N.S.F.

R. W. Hart, Assistant Director, Exploratory Development, The Johns Hopkins University, Applied Physics Laboratory

G. W. Ingle, Director, Association Liaison, Chemical Manufacturers Association

R. J. Kandel, Chief, Fundamental Interactions Branch Division of Chemical Sciences, D.O.E.

W. Klemperer, Assistant Director for Mathematical and Physical Sciences, N.S.F.

A. Kuwalsky, Program Director, Biophysics Program, N.S.F.

D. R. Lide, Jr., National Bureau of Standards

Hon. D. E. Mann, Assistant Secretary/Navy (Research, Engineering & Systems)

R. P. Mariella, Executive Director, A.C.S.

O. P. Manley, Office of Basic Energy Sciences, D.O.E.

R. NICHOLSON, Director, Chemistry Division, N.S.F.

G. S. OMENN, Associate Director, Office of Science and Technology Policy, White House

E. S. PIERCE, Director, Division of Chemical Sciences, D.O.E.

F. PRESS, Director, Office of Science and Technology Policy, White House

J. E. RALL, Director, Intramural Research Programs, National Institute of Arthritis, Metabolism and Digestive Diseases, N.I.H.

D. K. STEVENS, Director, Division of Materials Sciences, Basic Energy Studies, O.E.R.–D.O.E.

R. F. WATSON, Deputy Division Director, Division of Scientific Education, Development and Research, N.S.F.

M. K. WILSON, N.S.F.

M. C. WITTELS, Chief, Solid State Physics and Materials Chemistry, Office of Basic Energy Studies, O.E.R–D.O.E.

CONTENTS

Reports and Discussions

ASPECTS OF CHEMICAL EVOLUTION

XVIIth SOLVAY CONFERENCE ON CHEMISTRY

ADVANCES IN CHEMICAL PHYSICS
VOLUME LV

THE SOLVAY CONFERENCES AND THE INTERNATIONAL COOPERATION IN SCIENCE

JACQUES SOLVAY

Mr. President, Ladies, and Gentlemen: For us Belgians, it is a great occasion to be celebrating the one hundred fiftieth anniversary of the founding of Belgium as an independent state. It is particularly gratifying that the U.S. National Academy of Science and the International Solvay Institutes of Physics and Chemistry should mark this historic occasion by holding a joint symposium on "Aspects of Chemical Evolution." This is indicative of the traditional friendship that binds the Belgian and the American people in a community of common interests and goals.

As the great grandson of Ernest Solvay who founded the Institutes, it is a special privilege for me to be here to introduce the part of this event devoted to the history of the Solvay Institutes.

Ernest Solvay, industrial chemist and social reformer, was born in 1838, eight years after Belgian independence. He attended school as a boarder and read widely in chemistry and physics. Even at school he had converted his room into a chemical laboratory, and spent his vacations working on experiments.

The atmosphere at home was warm and inspiring, and his father imbued him with a sense of social responsibility.

He had hoped that Ernest would pursue a brilliant academic career in engineering at the University, but an attack of pleurisy forced him to give up plans for university studies. When 21 years old, Ernest accepted a job in a small gas factory because this position gave him the opportunity of satisfying his enthusiasm for chemical experimentation.

He conducted experiments that led to the ammonia process for producing sodium carbonate. On 15 April 1861, the eve of his twenty-third birthday, Ernest Solvay took out his first patent on "the industrial manufacture of carbonate of soda by means of sea salt, ammonia and carbonic acid."

1

Solvay and Company was formed as a family enterprise in 1863. Numerous difficulties were encountered at first until Solvay's ammonium process of manufacture was perfected, replacing the old Leblanc "black ash process."

Ernest Solvay immediately sought to give an international character to his enterprise. He established factories in France, Germany, England, United States, Austria, and Russia, and by the end of the 1870s they numbered more than 20. For Ernest Solvay, the pursuit of wealth was not an end in itself; it was a means to help society. He used the power of his wealth and prestige for the good of humanity by promoting social and scientific projects and institutions. He devoted much attention to educational and social problems and contributed to various philanthropic and research organizations.

Ernest Solvay regarded science, in its various forms, as the key that would open the door to a richer life for man. He had remarked once: "Among the new paths of science, I undertook to follow three directions; three problems which, in my view, form a certain unity. First, a general problem of physics—the constitution of matter in time and space; then a problem of physiology—the mechanism of life from its most humble manifestations up to the phenomena of thought processes; and finally, the third one, a problem complementary to the first two—the evolution of the individual and that of social groups."

Solvay sought to initiate this program in his life. In 1894 he endowed the Institut de Physiologie and the Institut des Sciences Sociales at the Free University of Brussels. He also endowed, in 1903, a School of Commerce, and a workman's educational center. In 1912 Ernest Solvay founded the International Institutes of Physics and Chemistry.

As I have mentioned, he was greatly interested in the problem of the constitution of matter in time and space. He was particularly attracted to the study of the structure of matter in the context of a theory of gravitation. By 1887 he had developed his own ideas on this subject.

In the spring of 1910 Solvay encountered Walther Nernst at the house of his collaborator Robert Goldschmidt in Brussels. Nernst was then a professor of chemistry in the University of Berlin and one of the most distinguished physical chemists of the day. At Goldschmidt's house, Solvay mentioned to Nernst his ideas on gravitation and the structure of matter, and wondered whether they could be brought to the attention of the great physicists. Those he had in mind were men like Planck, Lorentz, Poincaré, and Einstein. Solvay was also interested in the crisis that was developing in classical physics with the advent of relativity and quantum theories.

Solvay's desire to submit his work on the fundamental principles what he called "gravito-matérialitique" to the attention of Europe's leading physicists prompted Nernst to envision an international conference on the current problems of kinetic theory of matter and the quantum theory of radiation. The idea struck an immediate responsive chord in Solvay's mind, and he charged Nernst to explore it further with Planck, Lorentz, Einstein, and the other prominent physicists. Nernst was quick to pursue the idea immediately on his return from Brussels to Berlin.

Nernst and Planck had already discussed, early in 1910, the possibility of holding a conference dealing with the reform of the classical theory, although Planck had wondered whether the time was ripe and whether one should not wait another few years. Nernst's meeting with Ernest Solvay in Brussels gave fresh encouragement to holding the conference.

Hendrik Antoon Lorentz, the great Dutch physicist from the University of Leyden, was invited to assume the presidency of the conference. The acceptances came immediately, showing great enthusiasm. There was not a single refusal. On 27 September 1911, Ernest Solvay informed H. M. King Albert about the details of the planned scientific conference; the King immediately replied from the Château de Laeken, promising to "interest myself in these important scientific meetings, and follow attentively this occasion devoted to the higher science. . . ."

Allow me to express my gratitude for the support that the Belgian Royal Family has always given to the Solvay Institutes. We most regret that the King who has personally always graced the Solvay conferences by his presence cannot attend today as he had previously hoped.

The first "Conseil Solvay" took place in Brussels from 30 October to 3 November 1911. Its general theme was "The Theory of Radiation and the Quanta." The President of the Conference, as I have mentioned, was Lorentz, and the participants were from Germany: Nernst, Planck, Rubens, Sommerfeld, Warburg, and Wien; from England: Jeans and Rutherford; from France: Brillouin, Madame Curie, Langevin, Perrin, and Poincaré; from Austria: Einstein (he was in Prague then, which came under Austria), and Hasenöhrl from Vienna; from Holland: Kamerlingh Onnes; and from Denmark: Martin Knudsen. The scientific secretaries of the conference were Robert Goldschmidt of Brussels, Maurice de Broglie from Paris, and F. A. Lindemann from Berlin. The proceedings of the Conference were edited by Paul Langevin and Maurice de Broglie and published by Gauthiers-Villars, Paris, in 1912.

This was the first truly international conference of its kind. It was unique in its style and composition: an administrative council in Brussels; an international scientific committee to decide on a theme, rapporteurs, and participants, the goal being to further progress in a field ready for intense scrutiny. Such would become the pattern of all the following Solvay Conferences in Physics and Chemistry.

The Conference was immensely successful. It helped the general acceptance of quantum theory as a viable description of natural phenomena in the atomic and molecular domains, and it highlighted the problems that required solution, problems that would be attended to in the succeeding years and decades.

Encouraged by this success, Ernest Solvay, with the counsel and help of H. A. Lorentz, established a foundation on 1 May 1912, initially for a period of 30 years, to be called the *Institut International de Physique*, with the goal "to encourage research which would extend and deepen the knowledge of natural phenomena." The new foundation was intended to concentrate on "the progress of physics and physical chemistry," including the problems pertaining to them in other branches of the natural sciences. The inclusion of physical chemistry, as one of the fields to be encouraged by the Institute, was intended as a tribute to the role which Walther Nernst had played.

An International Scientific Committee was given responsibility for directing activities.

In 1913, after a series of exchanges with Wilhelm Ostwald and William Ramsay, Ernest Solvay established another foundation, the *Institut International de Chimie*, which embraced activity relating to chemistry. The two foundations were ultimately united into *Les Instituts Internationaux de Physique et de Chimie*, each one having its own Scientific Committee.

The scene was set for a new type of scientific meeting whereby a group composed of the most well-informed experts in a given field would meet to discuss problems at their frontier, and would seek to define steps for their solution. But for the interruptions caused by the two World Wars, these international conferences on physics and chemistry have taken place almost regularly since 1911 in Brussels. They have been exceptional occasions for physicists to discuss the fundamental problems which were at the center of interest at different periods, and have stimulated the development of physical science in many ways.

As with the Nobel Prizes, the great success and prestige of the Solvay Conferences in the scientific world derive from the fact that some of the greatest scientists became associated with them from the very beginning. Scientists like Lorentz, Planck, Einstein, J. J. Thomson, Poincaré, Rutherford, Bohr, Madame Curie, Langevin, William Pope, William Ramsay, and Walther Nernst had a tremendous impact on the development of physics and chemistry in the beginning of the century and they were all associated with the Solvay Conferences. Later on Louis de Broglie, Heisenberg, Schrödinger, Dirac, Pauli, Born, and many other great scientists continued to be associated with the activities of the International Solvay Institutes.

Einstein became a member of the Scientific Committee and took an active part in the discussions on quantum mechanics, which was the theme of the 1927 Conference. Indeed, the famous Einstein–Bohr discussions on classical determinism versus the quantum statistical causality took place in Brussels at the Solvay Conferences of 1927 and 1930, and continued thereafter.

Numerous participants in the various Solvay Conferences have told of the direct beneficial effects which they experienced on these occasions. Niels Bohr once said: "The careful recording of the reports and of the subsequent discussions at each of these meetings will in future be a most valuable source of information for students of the history of science wishing to gain an impression of the new problems raised in our century." Indeed, the gradual clarification of these problems through the combined efforts of a whole generation of physicists (and chemists) did, in the following decades, not only increase our insight in the atomic structure of matter, but even lead to a new outlook on the comprehension of physical experience.

"A new outlook on the comprehension of physical experience" is precisely what Ernest Solvay had been thinking when he convened the first Conference and then took the initiative to provide the forum and the means for the dialogue to go on.

It is a matter of personal satisfaction to me that my family has continued to be associated over the years with the functioning of the Solvay Institutes and their activities.

It is also most gratifying to follow the development of a scienfic institution, especially when its day-to-day management is conducted by a man of such outstanding and remarkable scientific achievements as is my friend Ilya Prigogine.

Today, 70 years after the first Solvay Conference, nature appears to be as mysterious as before, even more so. However, the emphasis has shifted in our times to diversification as well as to the impact of man on nature. It is therefore very timely that this conference is devoted to "Aspects of Chemical Evolution."

As President of the Solvay Foundation I wish to express my gratitude to the U.S. National Academy of Sciences, which made this occasion possible.

THE SOLVAY CONFERENCES
IN PHYSICS

EDOARDO AMALDI

Istituto di Fisica "Guglielmo Marconi,"
Universita degli Studi, Rome, Italy

I. INTRODUCTION

The number of Solvay Conferences in Physics, held in Brussels every few years, starting from 1911 until 1982, has been 17.

As it has been explained by Mr. Jacques Solvay, the first one had been conceived by Ernest Solvay in consultation with Walther Nernst, as a tool to help directly in solving a specific problem of unusual difficulty impending over the whole of physics: was the quantum structure of nature really unavoidable? Such a structure had been suggested by Max Planck[1] in 1900, for interpreting the observed spectrum of the blackbody radiation. Five years later Einstein[2] had again considered it in connection with the fluctuations of the energy density of the electromagnetic field and, as an example, he had applied it in an extended form, to the photoelectric effect. Einstein, had used it once more a few years later for the computation of the specific heat of solid matter.[3] But the majority of physicists disliked the introduction of such a procedure, which appeared in contrast with the whole conception of nature prevailing in those days.

In 1911 the idea of national and international conferences dealing with scientific matters, was about a century old. The first scientific conference, probably, was the one organized in 1815 in Geneva by the chemist H. A. Gosse on the "Physical and Natural Sciences." A number of scientific conferences had been held in many places during the following 96 years.

The first Solvay Conference in Physics, however, set the style for a new type of scientific meeting: as Mr. Jacques Solvay told us, the participants in the meeting formed a select group of the most informed experts in a given field which met to discuss one or a few related problems of fundamental importance and seek to define the steps for their solution.

This style was kept for many years although the extraordinary expansion undergone with the passing of time by the effort invested in the development of all the sciences, in particular of physics, has provided an

7

TABLE I. The Solvay Conferences in Physics

Inquire after laws in complex systems	Search for the last constituents of matter	Exploration of our environment at large
	1 Radiation Theory and 1911 the Quanta	
	2 The Structure of 1913 Matter	
	First World War	
	3 Atoms and Electrons 1921	
4 The Electrical 1924 Conductivity of Metals		
	5 Electrons and Photons 1927	
6 Magnetism 1930		
	7 The Structure of 1933 Atomic Nucleus	
	Second World War	
	8 Spectrum of 1948 Elementary Particles	
9 Solid State 1951		
10 Electrons in Metals 1954		
		11 The Structure and 1958 Evolution of the Universe
	12 Quantum Field 1961 Theory	
		13 The Structure and 1964 Evolution of Galaxies
	14 Fundamental 1967 Problems in Elementary Particles	
15 Symmetry 1970 Properties of Nuclei		
		16 Astrophysics and 1973 Gravitation
17 Order and 1978 Fluctuations in Equilibrium and Nonequilibrium Statistical Mechanics		

increasing number of other occasions for exchanges of views and discussions between experts in any subject.

The proceedings of the old Solvay Conferences in Physics forever will remain sources of information of a unique kind about the historical development of our present views. It is much more difficult, perhaps impossible, to make a similar statement about the more recent Solvay Conferences. Only in 10 or 20 years from now will they be seen in the right perspective. What we know with certainty, from now, is that the 17 Solvay Conferences in Physics provide in their ensemble a succession of pictures of the state of our knowledge of the physical world as it was changing at intervals of a few years, all taken with the "same camera" from essentially the "same point of view."

A global view of the Solvay Conferences in Physics can be grasped from Table I, where I have indicated the year in which each conference took place and its general theme. The conferences have been ordered in a rather oversimplified way into three columns. The central one contains the Solvay meetings aiming to search for the last constituents of matter and their properties, the column on the left those devoted to inquires about simple laws in complex systems, and that on the right the meetings concerned with the exploration of our environment at large.

Such a classification has been adopted here only to give an order to the presentation of the rather complex subject. The succession of these 17 conferences shows two interruptions due to the two World Wars, which lasted from 1914 to 1918 and from 1939 to 1945.

A volume by Jagdish Mehra, which appeared in 1975,[4] treats the subjects of these conferences and tries to indicate their scientific significance. It was of help in the preparation of my discussion.

II. THE CONFERENCES ON THE SEARCH OF THE LAST CONSTITUENTS OF MATTER

The participants in the 1st Solvay Conference in Physics are shown in Fig. 1, which reproduces the, perhaps, most famous photograph of physicists of any time.

Hendrik Antoon Lorentz, from Leyden (Holland), presided the conference, whose general theme was the "Theory of Radiation and the Quanta." The conference[5] was opened with speeches by Lorentz and Jeans, one on "Applications of the Energy Equipartition Theorem to Radiation," the other on the "Kinetic Theory of Specific Heat according to Maxwell and Boltzmann." In their talks, the authors explored the possibility of reconciling radiation theory with the principles of statistical mechanics within the classical frame. Lord Rayleigh, in a letter read to the

Photo Couprie, Bruxelles

GOLDSCHMIDT	PLANCK	RUBENS	LINDEMANN	HASENOHRL			
NERNST	BRILLOUIN	SOMMERFELD	DE BROGLIE	HOSTELET		
SOLVAY		KNUDSEN	HERZEN	JEANS	RUTHERFORD
LORENTZ	WARBURG	WIEN		EINSTEIN	LANGEVIN
PERRIN	Madame CURIE	POINCARÉ	KAMERLINGH ONNES

Fig. 1. Photograph of the participants in the First Solvay Conference in Physics (1911).

conference, stressed again the difficulty he had brought out in his masterly analysis[6] of 1900 and added: "Perhaps one could invoke this unsuccess as an argument in favour of the opinion of Planck and his school, that the laws of Dynamics (in their usual form) can not be applied to the last constituents of bodies. But I should confess that I do not like this solution of the difficulty. I do not see any inconvenience, of course, in trying to follow the consequences of the theory of the elements of energy (i.e., quanta). This method has already brought interesting consequences, due to the ability of those who have applied it. But it is difficult for me to consider it as providing an image of reality."

Two papers, one by E. Warburg and the other by H. Rubens, summarized the experimental measurements of the blackbody radiation for values of λT (λ in μm) smaller and larger than 3000.

These contributions were followed by an extensive presentation of the "Law of the Black Radiation" by Max Planck, who discussed, among other aspects, the physical nature of the constant h. Does this "quantum of action," he said, possess a physical meaning for the propagation of electromagnetic radiation in vacuum, or does it intervene only in the emission and absorption processes of radiation by matter?

The first point of view had been adopted by Einstein in the frame of his hypothesis of the "light quanta."[2]

Views of the second kind had been adopted by Larmor and Debye,[7] who conceived the quantum of action h as an elementary domain of finite extension in the space of phases intervening in the computation of the probability $W(E)$ for the energy density to have the value E.

In his contribution Nernst dealt with "the application of the Theory of Quanta to a few Physico-Chemical Problems," in particular the connection between Nernst theorem[8] and the quantization of energy.

Sommerfeld applied the theory of quanta to the emission of X- and γ-rays, to the photoelectric effect, and sketched the theory of the ionization potential. At the beginning of his paper he discussed in some detail the relationship observed in the emission of X- (or γ-) rays by cathode (or β) rays and arrived at the conclusion that "large quantities of energy are emitted in shorter times and small quantities of energy in larger times."[9] According to Sommerfeld this empirical result speaks in favor of the central role played in atomic and molecular phenomena by the quantum of action h introduced by Planck, the dimensions of which are energy multiplied by time.

The problem of specific heats, treated by Jeans from the classical point of view, as I said above, was discussed by Einstein in the case of solids, with special regard to the discrepancy observed at low temperature between the measured values and those deduced from the theory he had constructed in 1907 by quantizing the mechanical oscillators[3] as Planck had quantized the radiation oscillators.

Knudsen reported on the experimental properties of ideal gases, Kamerlingh Onnes on the electrical resistance of metals at low temperature, in particular on superconductivity he had discovered in Leiden in 1911,[10] and Langevin on the kinetic theory of magnetism and the central role played by the *magneton*, that is the magnetic moment of the elementary magnets which had been introduced in different approaches by Weiss and Langevin himself.

Finally Jean Perrin presented an extensive (97 pages) "Rapport sur les Preuves de la Réalité Moléculaire" in which he summarized his famous experiments on the Brownian motion of emulsion droplets suspended in a liquid and discussed the fluctuations, the determination of the elementary charge, the α decay of some radioactive nuclei, and the corresponding production of helium. The last section of the paper contains a comparison of the values of Avogadro's number deduced by completely different methods. The very satisfactory agreement between all these values provides the proof of molecular reality announced in the title of the paper.[11]

Nothing was said during the conference about the structure of the atom except in a short remark to Jeans' report by Rutherford in which he pointed out that the atom can be divided into two parts, an external part and an interior part and that the generalized coordinates, which according to Jeans appear not to contribute to the specific heat, could be those connected with the internal part of the atom.

The 2nd Solvay Conference (Fig. 2) took place in 1913 and its theme was "The Structure of Matter." [12] The meeting was opened with a long contribution (44 pages) by J. J. Thomson on the "Structure of the Atom" in which he tried to explain in a qualitative way from the classical point of view many general properties of matter.

In his long paper, however, there is no mention at all of the Rutherford model which appeared in the *Philosophical Magazine* of 1911[13] or of the papers by Geiger published in 1908 and 1910[14] and by Geiger and Marsden,[15] which appeared shortly before the meeting, on the scattering of α particles by atoms, or of the theoretical paper in which Niels Bohr had quantized the circular orbits of the electron of the Rutherford model.[16] Only in the discussion that followed J. J. Thomson's paper, did Rutherford mention the recent results of Geiger and Marsden which "bring to the conclusion that the atom consists of a positive nucleus, surrounded by a

Fig. 2. Photograph of the participants in the Second Solvay Conference in Physics (1913).

collection of electrons whose number is equal to one half the atomic weight. . . ." In a second intervention, in the same discussion, Rutherford added some more detail and said: "An accurate comparison of the theory with the experiments has been made by Geiger and Marsden and the conclusions of the theory have been found in perfect agreement with the experimental results."

Then Langevin pointed out that the central nucleus mentioned by Rutherford had to contain electrons in order to explain the emission of β-rays by radioactive atoms and Marie Curie elaborated the idea that within the atom there should be "two kinds" of electrons. The peripheral electrons responsible for the processes of absorption and emission of radiation and for conductivity of metals and, in addition, the electrons emitted in the β decay of some radioactive nuclei.

The successive report to the conference was presented by Marie Curie who, in 5 short pages, dealt with "The Fundamental Law of Radioactive Transformations." She discussed the exponential law of decay and its interpretation in terms of a probability for an atom to decay independently from its previous life. This point of view had been already recognized at the beginning of the century. However, on this occasion she added a short but adequate review of the idea of her pupil and collaborator Debierne[17] insisting on the existence of a "disorder inside the central part of the atom (the nucleus) where its constituents should move with very high velocity judging from that of the emitted particles." Madame Curie examined even the possibility of defining a kind of temperature internal to the nucleus, much higher than the external temperature.

The discussion that followed, in which Nernst, Rubens, Brillouin, Wien, Lundman, and Langevin took the floor, will remain forever one of the moments of highest interest in the history of the gradual infiltration of the probability law into physical sciences which foreran the advent of quantum mechanics and its statistical interpretation.[18]

The rest of the conference was dominated by the recent discovery of Friedrich, Knipping, and Laue[19] of the diffraction of X-rays in crystals and reviews of the first steps in X-ray spectroscopy and in the investigation of crystal properties.

The 3rd Solvay Conference in Physics took place in 1921, after a long interruption due to the First World War. Its theme was "Atoms and Electrons."[20] It was centered on the Rutherford model of the atom and Niels Bohr's atomic theory. Bohr, however, was not able to attend the conference because of illness.

In a speech, 20 pages long, Rutherford discussed "The structure of the Atoms." After a detailed presentation of the results of Geiger and Marsden on the scattering of α particles by atoms, Rutherford discussed the simple relation obtained by Moseley between the frequency of the X-ray lines of the different elements and their number of order in the periodic system,[21] which was then recognized to be identical with the ratio of the electric charge of the nucleus and the absolute value of the charge of the electron.

Rutherford gave an estimate of the dimensions of the nucleus and then discussed the passage of β and α particles through matter, the transmutation of nitrogen into oxygen he had observed 2 years before,[22] the existence of isotopes of both radioactive and stable elements, their separation, and finally the structure of nuclei for which a reduction of the mass had been established with respect to the sum of the masses of their constituents, which were assumed to be protons and electrons, in appropriate number.

Maurice de Broglie discussed the relation $E = h\nu$ in various phenomena, such as the photoelectric effect, the production of light and X-rays in collisions of electrons against atoms; Kamerlingh Onnes, the paramagnetism at low temperature and the superconductivity; and de Haas, the angular moment of a magnetized body. At the end of the conference Paul Ehrenfest read a paper sent by Niels Bohr on "The Application of the Theory of Quanta to Atomic Problems" and added a survey on "The Correspondence Principle." Also, a paper by Millikan on "The Arrangement and Movements of the Electrons Inside the Atoms" not presented to the conference was added at the end of the proceedings.

The next Solvay Conference along the same line of thinking was the 5th held in October 1927 on "Electrons and Photons."[23] Later Langevin said that this was the occasion in which "la confusion des idées atteignit son maximum!" Sir William L. Bragg commented "I think it has been the most memorable one which I have attended. . . ." Heisenberg and Bohr also expressed their highest satisfaction.[24]

Quantum mechanics had exploded between 1923 and 1927. A. H. Compton, in 1923, had discovered the change in frequency of X-rays scattered from the electrons (the Compton effect).[25] Compton and, independently, Debye had underlined the importance of this discovery in support of the Einstein conception of light-quanta or photon propagation in space.[26]

The paper of Louis de Broglie, associating a wavelength to any particle, had appeared in 1925.[27] The five notes of Schrödinger on wave mechanics appeared in 1926.[28] The various papers on the representation of physical quantities by matrices by Born, Heisenberg, Jordon, Dirac, and Pauli were published between 1925 and 1927.[29] The results of diffraction experiments by Davisson and Germer[30] of electrons scattered by a single crystal of nickel and by G. P. Thomsom and Reid[31] of electrons transmitted by celluloid films were in 1927.

The paper by Max Born on "Quantummechanik des Stossvorgänge," in which he had proposed the statistical interpretation of the wave function, had appeared in 1926.[32] Niels Bohr had presented his principle of complementarity at the Como Conference in September 1927[33] and Heisenberg had formulated the uncertainty principle shortly before the Solvay Conference.[34]

The Conference was opened on October 24 with reports by Lawrence Bragg and Arthur Compton about the new experimental evidence regarding scattering of X-rays by electrons exhibiting widely different features when firmly bound in crystalline structures of heavy substances and when practically free in atoms of light gases.

All other contributions regarded various aspects of the new quantum mechanics.

Louis de Broglie discussed the relations between energy, momentum, and wavelength for photons as well as for electrons and examined the results of the Davisson–Germer experiment on the diffraction of electrons by crystals, which were in perfect agreement with theory.

Born and Heisenberg presented the foundations of the quantum mechanics in which the classical kinematic and dynamical variables are replaced by operators obeying a noncommutative algebra involving Planck's constant and showed that the introduction of these operators in the Hamiltonian gives rise to the Schrödinger equation. Schrödinger summarized the main features of wave mechanics and applied it to radiation, showing that the electric moment deduced in this approach is equivalent to a matrix of the theory of Born and Heisenberg.

Finally, Niels Bohr gave a report on the epistemological problems involved in the new quantum mechanics and stressed the viewpoint of complementarity. In the very lively discussion that followed, differences in terminology gave rise to great difficulties for agreement between the participants. The situation was humorously expressed by Ehrenfest, who wrote on the blackboard the sentence from the Bible, describing the confusion of languages that disturbed the building of the Babel tower.

The discussions, started at the sessions, were continued during the evenings within smaller groups, in particular, among Bohr, Ehrenfest, and Einstein, who, as is well known, was reluctant to renounce the deterministic description.

The discussion on these matters between Einstein and Bohr went on for years. It was taken up again at the 1930 and 1933 Solvay Conferences as well as in other places. It has been summarized by Niels Bohr in a chapter of the well-known book *Albert Einstein: Philosopher-Scientist*, which appeared for the first time in 1949.[35]

At any occasion Einstein proposed a new argument or a new "thought experiment" designed to illuminate some paradoxical consequence of quantum mechanics. On some occasions immediately, on others some time later, Bohr always was able to give an answer fully satisfactory in the laguage of quantum mechanics. Bohr's arguments were accepted by the majority of the physicists but did not satisfy Einstein, or a few others[36] who still were reluctant to renounce a view of the world they considered obvious and natural. These world views, called by some authors "local realistic theories of nature,"[37] are based on three assumptions or premises which, in their opinion have the status of world established truths, or even self-evident truths.[38] This definition of "reality" is what we are used to ascribing to systems of macroscopic particles and differs from the "reality" of quantum states of atomic or subatomic systems.

The progress undergone in recent years toward a solution of the problem raised by Einstein 1927 has become possible because of two developments: (1) A few experimental techniques, in particular, the methods for measuring very short times with good accuracy, have permitted in recent years the execution of several experiments which in their essence are practical versions of the "thought experiment" proposed and discussed in 1935 by Einstein, Podolski, and Rosen.[39] (2) A procedure of analysis of their results has been made possible by the work of Bell[40] who has derived in the frame of "local realism" a relation (the Bell inequality) obeyed by "local realistic theories" but violated by quantum mechanics.

Although not all the results of the experiments carried on until now are consistent with one another, most of them not only violate the Bell inequality but are also in good agreement with the prediction of quantum mechanics. The experimental errors are still rather large but in a few years this matter will be definitely settled by the results of further, more accurate experiments. It is rewarding that the final word even in this philosophical debate started in 1927, when quantum mechanics was in its infancy, will be definitely settled by accurate results of experiments designed and carried on with the most refined techniques of half a century later.

The 7th Solvay Conference in Physics was held in 1933, and was entitled the "Structure and Properties of Atomic Nuclei."[41] It was chaired by Langevin who, after the death of Lorentz (4 February 1928), had become President of the Scientific Committee. It took place at a time when the field had recently undergone extremely rapid developments. The conference was really marking the beginning of modern nuclear physics.

Cockcroft presented an extensive paper (56 pages) on the "Disintegration of Elements by Accelerated Particles" in which he reviewed the various types of accelerators available in those days (Greinacher or Cock-croft and Walton voltage multiplier, van de Graaf electrostatic acceler-ator, cyclotron, etc.) and the reactions that he had discovered, in collab-oration with Walton, not long before in a few light elements (Li, F, Be, C, N) irradiated with protons and deuterons accelerated up to energies of the order of a few hundred kiloelectronvolts.[42]

In the discussion that followed, Rutherford added information about the work he was carrying on, in collaboration with Oliphant, on various reactions produced in lithium by bombardment with protons and deuter-ons,[43] and Ernst Lawrence described in more detail the cyclotron he had invented and first constructed with a few collaborators in 1932.[44]

The next speech was by Chadwick who treated (in 32 pages) the anom-alous scattering of the α particles, the transmutation of the elements, and the evidence for the existence of the "neutron" he had discovered in 1932.[45]

Frederick Joliot and Irene Curie discussed the γ-rays emitted in as-sociation with neutrons by berillium irradiated with α particles and re-ported to have observed under the same conditions also the emission of fast positrons. The origin of these particles was not yet clear at the time of the Solvay Conference. It was understood by the same authors a few months later when they discovered the artificial radioactivity induced by α-particle bombardment which normally takes place by emission of pos-itrons.

In the discussion that followed a number of participants took the floor: Lise Meitner, Werner Heisenberg, Enrico Fermi, Maurice de Broglie, Wolfgang Bothe, and Patrick Blackett.

In his intervention Blackett treated the discovery of the positron in cosmic rays by C. D. Anderson in 1932[46] and its confirmation by Blackett and Occhialini,[47] who had introduced, for the first time, the technique of triggering a vertical cloud chamber by means of the coincidence between two Geiger counters, one placed above, the other below the chamber. Blackett also discussed a number of papers by Meitner and Philipp, Curie-Joliot, Blackett, Chadwick and Occhialini, and Anderson and Nedder-meyer,[48] all appearing almost at the same time, on the production of positrons in various elements irradiated with the γ-rays of 2.62 MeV en-ergy of Thc. These were the first observations of electron–positron pair production. He also pointed out that the observed production of positrons has a cross section larger than the nuclear dimensions, and therefore, most probably, does not originate from a nuclear process.

Then Blackett examined the nature of showers observed in cosmic rays starting from Bruno Rossi's experiment with three nonaligned counters in coincidence and Rossi's curve showing the dependence of the number of observed showers as a function of the thickness of the layer of lead in which they are produced.[49] Then he presented the results he had recently obtained in collaboration with Occhialini by means of their new experimental technique mentioned above.[47]

The next paper was by Dirac on the "Theory of the Positron." In the following discussion Niels Bohr made a long intervention on the correspondence principle in connection with the relation between the classical theory of the electron and the new theory of Dirac.

Then Gamow talked about the "Origin of γ-Rays and the Nuclear Levels" and Heisenberg, on "The Structure of the Nucleus," discussed the exchange forces of the two types that Heisenberg[50] himself and Majorana[51] had proposed not long before.

In the discussion that followed, Pauli again brought up the suggestion he had already made in June 1931 on the occasion of a Conference in Pasadena. In order to explain the continuous spectrum of the β-rays emitted in the decay of many radioactive nuclei the emission of the electron had to be accompanied by the emission of a neutrino, that is, a neutral particle of a very small mass, possibly zero mass, and spin $\frac{1}{2}$.

Before the end of 1933 Fermi developed his theory of β decay,[52] in January 1934 Joliot-Curie discovered the radioactivity induced by α particles,[53] and Fermi that induced by neutrons.[54] Neutron physics was at the beginning of a new unexpected fast development.

Rutherford's comment to the 7th Solvay Conference was: "The last conference was the best of its kind that I have attended."

For a first time the theme selected for the 8th Conference was "Cosmic Ray and Nuclear Physics," but a long period of illness of the President of the Scientific Committee, Paul Langevin, imposed a first adjournment. Later it was decided that the conference would deal with the problems of elementary particles and their mutual interactions and that it would be held in October 1939. Even the list of speakers was prepared but World War II started on 3 September 1939 and the conference was postponed to an indefinite date.

The next Solvay Conference in Physics, the 8th from the beginning, took place in October 1948, that is, 15 years since the previous one. The Scientific Committee was now chaired by Sir Lawrence Bragg. The general theme of the conference was "Elementary Particles."[55]

Already in 1936 Anderson and Neddermeyer[56] had shown that the penetrating component of cosmic rays, first discovered in 1932 by Bruno Rossi,[57] consisted of charged particle of mass intermediate between those of the electron and the proton. Shortly later it was shown that this particle was unstable with a mean life on the order of 10^{-6} sec.

Both the mass and the mean life of this particle were in qualitative agreement with those suggested about 1 year before by Yukawa for the mediator of the nuclear forces.[58]

The picture, however, had changed rapidly after the end of the Second World War. The experiment of Conversi, Pancini, and Piccioni[59] had shown that this particle had an interaction with nuclei much weaker than that expected for the Yukawa mediator. At the beginning of October of the same year 1947, Lattes, Occhialini, and Powell[60] in Bristol had discovered in cosmic rays a new particle, that they called π-meson. It is unstable and decays, with a mean life of $\sim 10^{-8}$ sec, into a neutrino and the particle of Anderson and Neddermeyer that was called μ-meson or muon.

Almost at the same time Rochester and Butler[61] at Manchester observed in a cloud chamber triggered by counters two "V events" identified later as decays of a θ^0 ($\equiv K^0$)-meson and a Λ^0-hyperon.

These were the first examples of "strange particles," the adjective "strange" referring to the following anomalous property. Their decay takes a time of the order of 10^{-10} sec instead of the much shorter time ($\sim 10^{-23}$ sec) expected from the time observed to be involved in their production ($\sim 10^{-23}$ sec).

In Brussels the 8th Solvay Conference was opened with two speeches on mesons, one by Cecil F. Powell who reported on the results of work made in cosmic rays, and the other by R. Serber who presented the results on the production of "artificial mesons" obtained in Berkeley by Burfening, Gardner, and Lattes by bombarding matter with α particles accelerated with the cyclotron.[62]

The properties of the particles contained in large atmospheric showers were discussed by Auger and the problem of the nuclear forces by Rosenfeld from two points of view. In a purely phenomenological approach by means of a nuclear potential or in terms of a nuclear field mediated between any two interacting particles by particles of intermediate mass.

Bhabha treated the general relativistic wave equations and Tonnelat presented the idea of Louis de Broglie of trying to describe a photon as an object composed of two neutrinos.[63]

Heitler spoke on the quantum theory of damping, which is a heuristic attempt to eliminate the infinities of quantum field theory in a relativistic invariant manner, Peierls spoke of the problem of self-energy, and Oppenheimer gave "an account of the developments of the last years in electrodynamics" in which he discussed the problem of the vacuum polarization and charge renormalization with special reference to the recent work of Schwinger and Tomonaga.

A few reports on different topics were also presented to the same conference. Blackett discussed "The Magnetic Field of Massive Rotating

Bodies'' and Teller presented a paper prepared jointly with Maria Goeppert Mayer, on the original formation of the elements.

The conference was closed with some general comments by Niels Bohr ''on the present state of atomic physics'' in which he referred to the satisfactory situation in quantum electrodynamics, but pointed out that the adimensional coupling constant

$$\alpha = \frac{e^2}{\hbar c} \simeq \frac{1}{137}$$

cannot be computed within the framework of the theory itself, indicating the need for a future more comprehensive theory.

The 12th Solvay Conference on ''Quantum Field Theory'' was held in 1961, just 50 years after the first Solvay Conference.[64] This circumstance was celebrated by Niels Bohr, who opened the conference with a speech on ''The Solvay Meetings and the Development of Quantum Physics.''

Once more the development undergone between 1948 and 1961 by the experimental techniques as well as by the theoretical interpretation of subnuclear particles was amazing. In 1952, at the Brookhaven National Laboratory, the first proton synchrotron, the Cosmotron, entered into operation. It produced protons of energies up to 3.2 GeV, and became immediately a controlled source of pions and strange particles of much higher intensity than cosmic rays.

In the same year, Pais[65] suggested that the anomalous property of strange particles could be explained by assuming that they possess an internal degree of freedom, specified by a quantum number, and that various selection rules based on the conservation or nonconservation of this quantum number are operating in the production and decay.

Within 1 year, at the Cosmotron, Fowler et al.[66] discovered a new phenomenon, the associated production in the same collision of two different strange particles: for example, a Λ^0 and a K^0. These observations were in agreement with Pais' approach and prompted its full development: in 1954 Gell-Mann and Pais and, independently, Nishijima[67] were able to define the new quantum number, called strangeness, and, from a detailed analysis of the already rich harvest of new processes, to assign its value to all known particles.

In 1954, a larger accelerator, the Bevatron, producing protons up to 6.5 GeV, went into operation at Berkeley, and marked the end of cosmic rays as a tool for the investigation of subnuclear particles.[68] About 1 year later with this machine, Chaimberlain, Segrè, Wiegand, and Ypsilantis[69] observed the production of antiprotons generated in proton–nucleon collisions. This result provided the long-awaited confirmation of the generality of the 1931 Dirac forecast: his relativistic equation providing an adequate description of the electron and its antiparticle, is valid in general (apart from some correction) for any particle of spin $\frac{1}{2}$.

A further step of fundamental importance was made in 1961 by Gell-Mann and Ne'eman who proposed a classification scheme based on the Sophus Lie group of symmetry SU(3)[70] for the more than 100 particles endowed with strong interactions (hadrons).

About 2 years before the 1961 Solvay Conference two new accelerators producing protons of 28–30 GeV went into operation, one at CERN, near Geneva, and the other at the Brookhaven National Laboratory. Pretty soon, both of them started to enrich even more our knowledge of sub-nuclear phenomenology.

At the 1961 Solvay Conference the situation reached by quantum field theory was reviewed by Heitler, with special regard to the problem of renormalization. The successive speech on the same general theme was given by Feynman who started with a comparison of the predictions of the quantum field theory with experimental results. He discussed the scattering of high-energy electrons by protons, with special emphasis on the proton form factors and the search for deviations from the photon propagator, the comparison between the measured and computed values of the Lamb shift, of the magnetic moment of electron and muon, and of the hyperfine structure of spectral lines.[71] As a general conclusion on the first part of his talk, Feynman stated: "All this may be summarized by saying that no error in the prediction of quantum electrodynamics has yet been found. The contributions expected from various processes envisaged have been found again, and there is very little doubt that in the low-energy region, at least, our methods of calculation seem adequate today." These remarks are valid even today in spite of the fact that the precision reached in the measurements in the low-energy experiment is increased by two or three orders of magnitude[72] and the exploration of high-energy phenomena has been extended from values of the cut-off parameters Λ of 0.6 GeV to more than 100 GeV.[73]

The second part of Feynman's speech dealt with theoretical questions. The first one was the problem of the renormalization of the mass of the electron as well as of particles such as the pion and the kaon which exist in charged ($\pi^{\pm}$, $K^{\pm}$) and neutral (π^0, K^0) states and therefore provide a direct indication of the contribution originating from the electromagnetic field.

A second question was the interaction of photons with other particles which do not allow a sharp separation between quantum electrodynamics and other interactions. "It will not do to say that Q.E.D. is exactly right as it stands, because virtual states of charged baryons must have an influence. Two sufficiently energetic photons colliding will not do just what Q.E.D. in that limited sense supposes; they also produce pions. Or, more subtly, a sufficiently accurate analysis of the energy levels of positronium would fail, for the vacuum polarization from mesons and nucleons would be omitted."

A third and last theoretical item treated by Feynman was "Dispersion Theory," discussed in greater detail by other participants in the Solvay Conference.

"Weak Interactions" were treated by Pais who, starting from Fermi's original theory, discussed the discovery by Lee and Yang,[74] almost 5 years before, of the parity violation by weak interactions, its experimental confirmation,[75] the muon–electron universality,[76] the idea of an intermediate boson as a mediator of weak interaction, and the "two-neutrinos question."[77]

In the following speech, on "Symmetry Properties of Fields," Gell-Mann discussed exact symmetries as well as approximate symmetries: the conservation of the x and y component of the isotopic spin I, broken by electromagnetism and weak interactions, the conservation of I_z, or strangeness, broken by weak interactions and the conservation of C and P separately, also broken by weak interactions. Then, using various arguments, among them the conserved vector current, already recognized in 1955–1958,[78] he stressed the central interest of charge operators and of the equal-time commutation relations among them. "The mathematical character of the algebra which the charge operators generate is a definite property of nature," Gell-Mann said.

He then considered the currents in the Sakata–Okun model for which he derived the expressions for the electromagnetic and weak currents.

At this primitive stage of development of current algebra quarks had not yet appeared on the scene.

Other speeches were by Källen on some aspects of the formalism of field theories, Goldberger on single variable dispersion relation, Mandelstam on "Two-Dimensional Representations of Scattering Amplitudes and Their Application" and finally by Yukawa on "Extensions and Modifications of Quantum Field Theory."

The appearance of the current algebra in the speeches and discussions of this conference really marks the beginning of one of the most fruitful lines of development that has brought us to present views.

The 14th Solvay Conference on "Fundamental Problems in Particle Physics" was held in October 1967.[79] The Conference was presided over by Christian Møller who opened the meeting with a "Homage to Robert Oppenheimer," who had died on 18 February of that year.

Many speeches were of a theoretical nature: Dürr spoke on "Goldstone

Theorem and Possible Applications to Elementary Particle Physics," Haag on "Mathematical Aspects of Quantum Field Theory," Källen on "Different Approaches to Field Theory. Especially Quantum Electrodynamics," and Sudarshan on "Indefinite Metric and Nonlocal Field Theories." Heisenberg gave a "Report on the Present Situation in the Nonlinear Spinor Theory of Elementary Particles."

Chew spoke of the "S-Matrix Theory with Regge Poles," a concept[80] that soon became a stable acquisition in the general picture of subnuclear particles,[81] and Tavkhelidze discussed the "Simplest Dynamic Models of Composite Particles."

Already in 1964 Zweig and Gell-Mann had postulated the existence of quarks as building blocks of hadrons, thus establishing the premises necessary for a dynamical interpretation of the SU(3) already well-established symmetry.[82] At the time of the Solvay conference, however, the quark hypothesis was considered only as a convenient model.

A report by Gell-Mann, also on elementary particles, unfortunately does not appear in the proceedings of the Conference because it was not available at the time of publication. From the contributions to its discussion by W. Heisenberg, C. F. Chew, F. E. Low, S. Mandelstan, R. Brout, R. E. Marshak, S. Weinberg, N. Cabibbo, S. Fubini and others, it appears that the talk by Gell-Mann touched upon the SU_2 and SU_3 groups and their connections with bootstrap mechanism, chiral symmetry, and partial conservation of axial currents (PCAC).[83]

This is the last of the Solvay Conferences on Physics devoted to subnuclear particles in spite of the fact that the progress undergone by this field since 1967 has been amazing. The main reason was that many conferences were held each year on this general subject as well as on various parts of it, so that the Scientific Committee for Physics felt it would be better to turn its attention to other subjects less frequently dealt with in international conferences.

Now it is clearly time to devote yet another Solvay Conference to the last constituents of matter. See note on page 35.

III. THE CONFERENCES ON THE INQUIRE AFTER LAWS IN COMPLEX SYSTEMS

Let me now consider the conferences devoted to solid-state and statistical mechanics.

The first conference of this group is the 4th Solvay Conference on "The Electrical Conductivity of Metals" held in April 1924.[84] The conference was in some way premature. It took place just before the advent of quantum mechanics, in particular 2 years in advance of the first formulation by Fermi of the antisymmetric statistics and the consequent concept of the degenerate electron gas.

The conference was opened with a speech by Lorentz on the theory of electrons he had developed about 20 years before, followed by papers by Joffe on the electrical conductivity of crystals, Kamerlingh Onnes on superconductivity, and Hall on the metallic conduction and the transversal effects of the magnetic field. This last speech was followed by a discussion in which Langevin and Bridgman injected a few interesting remarks.

The 6th Solvay Conference on "Magnetism," held in 1930,[85] was opened by a contribution by Sommerfeld on "Magnetism and Spectroscopy" in which he discussed the angular momenta and magnetic moments of the atoms which had been derived from the investigation of their electronic constitution.

Van Vleck reported on the experimental data of the variation of the magnetic moments within the group of the rare earths and its theoretical interpretation, and Fermi discussed the magnetic moments of the atomic nuclei and their determination from the splitting of hyperfine structure. Pauli treated the "Quantum Theory of Magnetism," with special regard to the paramagnetism of a degenerate Fermi gas of electrons, Weiss dealt with the equation of state of ferromagnets, Dorfman with ferroelectric materials, and Cotton and Kapitza reported on the study of the magnetic properties of various materials in very intensive magnetic fields. The phenomenological theory of ferromagnetism by Weiss is still of interest today, mainly because the microscopic mechanism that gives rise to the dipole–dipole interaction is not yet understood in all its detail.[86]

The 9th Solvay Conference on "Solid State" took place in 1951.[87] The question of "interface between crystals" was discussed by C. S. Smith, grain growth observed by electron optical means by G. W. Ratenau, recrystallization and grain growth by W. G. Burgers, crystal growth and dislocation by F. C. Franck, the generation of vacancies by moving dislocations by Seitz, dislocation models of grain boundaries by Shockley, and diffusion, work-hardening, recovery, and creep by Mott.

It was at that time that Franck and Seitz proposed mechanisms for the multiplication and generation of vacancies by intersection of dislocations[88] explaining the observed softness of crystals and providing models that were subsequently verified by the technique of decoration of dislocations.[89]

The 10th Solvay Conference on "The Electrons in Metals" was held 3 years later, in 1954.[90]

D. Pines examined the collective description of electron interaction in metals; Löwdin, an extension of the Hartree–Fock method to include correlation effects; Mendelson, the experiments on thermal conductivity of metals; Pippard, the methods for determining the Fermi surface; Kittel, resonance experiments and wave functions of electrons in metals; Friedel, primary solid solutions in metals; Fumi, the creation and motion of vacancies in metals; Shull, neutron diffraction from transition elements and their alloys; Néel, antiferromagnetism and metamagnetism; and Frölich, superconductivity with special regard to the electron–electron interaction carried by the field of lattice displacements, which paved the way to the theory that Bardeen, Cooper, and Schrieffer developed between 1955 and 1957.[91] Finally, Mathias dealt with the empirical relation between superconductivity and the number of valence electrons per atom.

The last Solvay Conference that I have rather arbitrarily put in this class is the 17th on "Order and Fluctuations in Equilibrium and Nonequilibrium Statistical Mechanics" held in October 1978.[92]

The conference was divided into four parts to each of which a full day was devoted: the first one treated: "Equilibrium Statistical Mechanics," with special regard to "The Theory of Critical Phenomena"; the second part regarded "Nonequilibrium Statistical Mechanics. Cooperative Phenomena"; the third one, "The Macroscopic Approach to Coherent Behavior in Far Equilibrium Conditions"; and the fourth and last, "Fluctuation Theory and Nonequilibrium Phase Transitions."

Two methods appear to be very powerful for the study of critical phenomena: field theory as a description of many-body systems, and cell methods grouping together sets of neighboring sites and describing them by an effective Hamiltonian. Both methods are based on the old idea that the relevant scale of critical phenomena is much larger than the interatomic distance and this leads to the notion of scale invariance and to the statistical applications of the renormalization group technique.[93]

As pointed out by van Hove in his concluding remarks, the common methodology between high-energy physics and critical phenomena is striking although the conceptual basis is quite different in the two cases. In high-energy physics approximate scale invariance and its calculable breaking are characteristic of the large momentum scale regime (i.e., small space and time intervals), whereas in statistical physics scale invariance and renormalization group methods are applicable to a domain of large space and time intervals.

The approximate methods of renormalization for the investigation of phase transitions in degenerate states[94] were presented to the conference by Kadanoff and by Brezin. The nonequilibrium statistical methods were discussed by Prigogine,[95] followed by Hohenberg who treated critical dynamics. In the third part, Koschmieder discussed the experimental aspects of hydrodynamic instabilities[96]; Arecchi, the experimental aspects of transition phenomena in quantum optics[97]; and Sattinger the bifurcation theory and transition phenomena in physics.[98]

The fourth part included a paper by Graham on the onset of cooperative behavior in nonequilibrium states. Suzuki talked about the theory of instability, with special regard to nonlinear Brownian motion and the formation of macroscopic order, and P. W. Anderson developed a series of interesting considerations of very general nature around the question: "Can broken symmetry occur in driven systems?"

The question in the title can be reformulated by asking how much can be dug out of an analogy between broken symmetry in dissipative structures (such as the ripple marks generated by wind, i.e., an external perturbation, in an otherwise flat surface of sand) and broken symmetry defined as phenomena of condensed matter systems of the kind observed near the critical points. The value of Anderson's discussion is to be seen more in the deepening of the question itself than in the answer that cannot yet be final, and for the moment, according to the author, appears to be more on the negative side.

The contributions presented by Prigogine and by Sattinger to the 17th Solvay Conference on Physics appear as a natural introduction to some of the problems that will be examined at the present Solvay Conference in Chemistry.

Exact symmetries and broken symmetries were the central theme of the 15th Solvay Conference held 8 years before, that is, in 1970 on "Symmetry Properties of Nuclei."[99] This was the second conference on nuclear structure, the previous one being the conference held in 1933 immediately after the discovery of the neutron. In the 37 years that have passed between the first and the second Solvay Conference on nuclear structure the subject has undergone an extraordinary development although the naive hope, generally shared by physicists until 1935, for a full understanding of nuclear dynamics in terms of nucleons interacting with two-body forces, has been completely deluded. A number of models have been developed which, although very different from each other, are not contradictory. Each of them, in some way, enphasizes a particular aspect of some category of nuclei and/or nuclear phenomena, and thus allows an adequate interpretation of a set of their static and/or dynamic properties.

The "Symmetry of Cluster Structures of Nuclei" was discussed by Brink who showed contour plots of nucleon density obtained from Hartree–Fock calculations for simple nuclei such as ^8Be, ^{12}C, and ^{20}Ne.

Much attention was devoted to collective models: Mottelson reviewed "Vibrational Motion in Nuclei"; Judd, the use of Lie groups; Lipkin, the

SU(3) symmetry in hypernuclear physics; Radicati, Wigner's supermultiplet theory[100]; Fraunfelder, "Parity and Time Reversal in Nuclear Physics"; Wilkinson, the isobaric analogue symmetry; Aage Bohr, the permutation group in light nuclei; and J. P. Elliot, the shell model symmetry.

The conference was closed by a few concluding remarks by E. P. Wigner, not completely free from critical lines.

At the time of the conference the study of nuclear reactions produced by intermediate energy protons or pions as well as the investigation of collisions between two nuclei of $Z \geq 3$ were still in a rather primitive stage whereas today they constitute a rich field of empirical knowledge and phenomenological interpretation.

IV. THE CONFERENCES ON EXPLORATION OF OUR ENVIRONMENT AT LARGE

I come now to the last group of Solvay Conferences: the three regarding astrophysical problems. The first one, devoted to "The Structure and Evolution of the Universe" was held in June 1958. It was the 11th Solvay Conference in Physics.[101]

The conference took place in a moment of extraordinary expansion of general interest for astrophysical problems due to the first steps made in new observational techniques such as radio signal reception and observations from space vehicles.

This was also the first Solvay Conference in which Einstein's Theory of General Relativity started to be quoted and used as a conceptual structure of fundamental importance for the interpretation of large-scale phenomena.

The theme of the conference was divided into three parts: the first one concerned "General Statements of Cosmological Theory." It was introduced by speeches by Lemaitre, on the "Primaeval Atom Hypothesis and the Problem of Clusters of Galaxies," by Oscar Klein who developed "Some Considerations Regarding the Earlier Development of the System of Galaxies," and by Hoyle on "The Steady-State Theory." This was followed by a talk by Gold, on the "Arrow of Time" and another by Wheeler on "Some Implications of General Relativity for the Structure and Evolution of the Universe."

The second part of the conference was devoted to a "Survey of Experimental Data on the Universe."

In a talk, very important even today, J. H. Oort discussed the "Distribution of Galaxies and the Density of the Universe." Lovell presented "Radio Astronomical Observations Which May Give Information on the Structure of the Universe."

In the third part of the conference, on the "Evolution of Galaxies and Stars," Hoyle presented the then recent and still important work by the Burbidges, Fowler, and himself on the origin of elements in stars.[102]

In the general discussion Bondi asked for tests that could decide between the evolutionary and steady-state universe. The question was premature because the arguments, which later allowed the exclusion of the steady-state theory, are based on the analysis of the blackbody cosmological radiation and of the distribution of the number of radio sources versus flux. The cosmological radiation was discovered only in 1965[103] and the distribution of the radio sources was still highly controversial at the time of the conference. Indeed, in the discussion of this point Lovell stated: "At present the whole of the cosmological interpretation of the radio sources is based on a half a dozen identifications of the Cygnus type." Furthermore, the mechanism leading to radio emission was misunderstood: the prevailing view attributed it to collisions between galaxies.

From the small amount of information I have given it appears clear that in 1958 astrophysics was still in its classical stage. The conference, however, marks a historical step in modern astrophysics because for the first time the physics of neutron stars and collapsed objects was reproposed by Wheeler since 1939 when Oppenheimer and Snyder first presented the existence of black holes.[104]

The 13th Conference on the "Structure and Evolution of Galaxies" took place in October 1964.[105] It was presided over by Robert Oppenheimer, who had succeeded Sir Lawrence Bragg.

The Conference was opened by Ambartsumian who spoke "On the Nuclei of Galaxies and their Activity." He presented new observational data that in his opinion supported the idea, which he had already submitted to the 1958 Conference, that most of the processes connected with the formation of new galaxies and their structure start from the nuclei. This interpretation, however, is shared only by a minority of astrophysicists.

Oort reported on "Some Topics Governing the Structure and Evolution of Galaxies" and described the latest knowledge concerning our Galaxy. Woltjer discussed the "Galactic Magnetic Field."

The structure and evolution of the stars were the subject of the second part. Spitzer discussed the "Physical Processes in Star Formation," a subject that was further developed by Salpeter with special regard to the birthrate function of the stars. W. A. Fowler and Bierman discussed the evolution toward the main sequence and R. Minkowski discussed the data available concerning the supernovae.

Finally, an important survey of the findings about extragalactic radio sources was presented by J. G. Bolton and, in the discussion that followed, Bruno Rossi presented the first observations made by Giacconi et al.[106] in 1962 and by Friedman et al. in 1963 of localized X-ray sources, in particular, Scorpio X-1.[107]

One of the most exciting contributions to the conference was the speech by M. Schmidt who discussed the "Spectroscopic Observations of Extragalactic Radio Sources," with special regard to the interpretation of the red shift of the quasars, discovered about 2 years before.[108] His conclusion, still accepted today by the majority of astrophysicists, was that most likely these red shifts are cosmological.

G. R. Burbidge and E. M. Burbidge, in their report on "Theories and the Origin of Radio Sources," summarized the understanding that had been achieved in the creation of radio waves.

In the final discussion G. R. Burbidge stressed that the problems of energy conversion from a primary energy source to the form of relativistic particles needed for the radio emission are entirely unsolved. In particular, the production of cosmic ray particles requires an acceleration mechanism, of an efficiency at least a few orders of magnitude larger than that of the best man-made accelerators.[109]

In this connection Alfvèn proposed the annihilation of matter and antimatter as a possible source of energy, but also other mechanisms, in particular, some form of release of gravitational energy were examined.

The problem is still open today, but the consensus is indeed that strong gravitational fields should be involved.

The 16th Solvay Conference in Physics, held in September 1973, was entitled "Astrophysics and Gravitation."[110] The progress undergone in many fundamental chapters of astrophysics with respect to the previous conference, held in 1964, was really striking. In particular, X-ray astronomy had won a status comparable with other conventional branches of astronomy.

The Conference was opened with a progress report on pulsars by Pacini, followed by speeches on the observational results on compact galactic X-ray sources by Giacconi, on the optical properties of binary X-ray sources by the Bahcalls, and a review on the physics of binary X-ray sources by Martin Rees.

Among the many communications and invited talks essentially on the same subject, I recall the speech by Pines on "Observing Neutron Stars," by Pandharipande on "Physics of High Density and Nuclear Matter," and by Cameron and Canuto on "The General Review on Neutron Stars Computations."

Black holes were extensively discussed by J. A. Wheeler and the search for their observational evidence by Novikov, followed by communications by Rees and by Ruffini.

Woltjer gave a general talk on "Theories of Quasars," and Martin Schmidt discussed "The Distribution of Quasars in the Universe."

G. R. Burbidge gave a review paper on the "Masses of Galaxies and the Mass-Energy in the Universe" and Hofstadter presented the information available at the time on the recent discovery of bursts of γ-rays.[111]

I am now at the end of my series of flashes on the Solvay Conferences in Physics. I hope that, in spite of its shortness and incompleteness, it may help in stimulating two kinds of considerations. Those of the first kind regard the extraordinary develoment undergone during the last 70 years by our views on the physical world, many parts of which in present days appear to be dominated by a few general concepts, such as those of exact and approximate symmetry, and to be treatable by mathematical procedures such as the application of the renormalization group. The other kind of considerations concerns the role that the Solvay Conferences in Physics have played in the development of physics during the last 70 years, and the unique value they will maintain, even in the future, as sources of information for the historians of science.

References

1. M. Planck, *Verh. Deut. Phys. Ges.*, **2**, 237 (1900).

2. A. Einstein, *Ann. Phys. Ser. 4*, **17**, 132 (1905); **20**, 199 (1906).

3. A. Einstein, *Ann. Phys.*, **22**, 180, 800 (1907); **25**, 679 (1911). Debye developed his model about one year later: *Ann. Phys.*, **39**, 789 (1912).

4. J. Mehra, *The Solvay Conferences on Physics*, D. Reidel, Dordrecht and Boston, 1975.

5. *La Théorie du Rayonnement et les Quanta*, Rapports et Discussions de la Rénunion tenu à Bruxelles, du 30 October au 3 Novembre 1911, Publiés par MM. Langevin et M. de Broglie, Gauthier-Villars, Paris, 1912.

6. Lord Raleigh, *Philos. Mag.*, **49**, 118 (1900); **59**, 539 (1900).

7. J. Larmor, *Proc. R. Soc. London, Ser. A*, **83**, 82 (1909); P. Debye, *Ann. Phys.*, **33**, 1427 (1910)

8. W. Nernst, Göttinger Nachr. 1906 Heft 1, *Zeit. Elektrochemie*, **17**, 265 (1911).

9. The argument goes more or less as follows: high-energy cathode rays, that have a short interaction time with atoms, emit X-rays of an energy (deduced from their penetration) greater than the energy of the X-rays emitted by cathode rays of lower energy. The conclusion is correct, but it is based on the presumption that the penetration of photons is a monotonic increasing function of their energy. This is true only for photons of energy below the threshold for pair production: $2m_e c^2 \simeq 1$ MeV. As I will recall below, pair production was discovered only in 1932, and anyhow the X-ray photons available and considered in 1911 had always energy below $2m_e c^2$.

10. K. Onnes, *Comm. Phys. Lab. Univ. Leyden*, Nos. 119, 120, 122 (1911).

11. *Oeuvres Scientifiques de Jean Perrin*, CNRS 13, Quai Anatole France, Paris (VIIe), 1950.

12. *La structure de la matiere*, Rapports et discussions du Conseil de Physique tenu à Bruxelles du 27 au 31 October 1913, Gauthier-Villars, Paris, 1921.

13. E. Rutherford, *Philos. Mag.*, **21**, 669 (1911).

14. E. Geiger, *Proc. R. Soc. London, Ser. A*, **81**, 174 (1908); E. Marsden, *Proc. R. Soc. London, Ser. A*, **82**, 495 (1909); E. Geiger, *Proc. R. Soc. London, Ser. A*, **83**, 492 (1910).

15. H. Geiger and E. Marsden, *Philos. Mag.*, **25**, 604 (1913).

16. N. Bohr: "On the Constitution of Atoms and Molecules", *Philos. Mag.*, **26**, 1913, p. 1 of July issue; p. 476 of September issue; p. 857 of November issue.

17. A. Debierne, "Sur les transformations radioactives", p. 304 of the volume: *Les idées modernes sur la constitution de la matiere*, Gauthier-Villars, Paris, 1913.

18. E. Amaldi, "Radioactivity, a Pragmatic Pillar of Probabilistic Conceptions," p. 1 of the volume: *Problems in the Foundations of Physics*, edited by G. Toraldo di Francia, North-Holland, Amsterdam, New York, Oxford, 1979.

19. W. Friedrich, P. Knipping, and M. v. Laue, *Ber. Byer. Akad. Wiss.*, 303 (1912).

20. *Atoms et Électrons*, Rapports et Discussions du Conseil de Physique tenu à Bruxelles du 1er au 6 Avril 1921, Gauthier-Villars, Paris, 1923.

21. H. G. J. Moseley, *Philos. Mag*, **26**, 1024 (1913); **27**, 103 (1914).

22. E. Rutherford, *Philos. Mag.*, **37**, 581 (1919).

23. *Électrons et Photons*, Rapports et Discussions du Cinquiènne Conseil de Physique tenu à Bruxelles du 24 au 29 October 1927, Gauthier-Villars, Paris, 1928.

24. These remarks can be found in the few pages of introduction to the album of photographs of the participants in the Solvay Conferences in Physics, published in 1961, on occasion of the celebration of 50 years after the First Solvay Conference. These pages are based on the unpublished paper: Jean Pelseneer, "Historique des Instituts Internationaux de Physique et de Chimie Solvay, depuis leur fondation jusqu' à la deuxiéme guerre modiale," Bruxelles, 1946.

25. A. H. Compton, *Bull. Nat. Res. Council* **XX**, 16 October 1922; *Philos, Mag.*, **46**, 897 (1923).

26. A. H. Compton, *Phys. Rev.*, **22**, 409 (1923); P. Debye, *Phys. Zeit.*, **24**, 161 (1923).

27. L. de Broglie, *Nature (London)*, **112**, 540 (1923); Thése, Paris (1924); *Ann. Phys.* 8(10), 22 (1925).

28. E. Schrödinger: *Ann. Phys.* 79(4), 361 (1926); *Naturwissenschatten*, **14**, 664 (1926); *Ann. Phys.* 79(4), 734 (1926); *Ann.Phys.* 80(4), 437 (1926); *Ann. Phys.* 81(4), 109 (1926).

29. W. Heisenberg, *Z. Phys.*, **33**, 879 (1925); M. Born and P. Jordon, *Z. Phys.*, **34**, 858 (1925); P. Dirac., *Proc. R. Soc. London, Ser. A*, **109**, 642 (1925), **110**, 561 (1926); M. Born, W. Heisenberg, and P. Jordan, *Z. Phys.*, **35**, 557 (1926); W. Pauli, *Z. Phys.*, **36**, 336 (1926); W. Heisenberg and P. Jordan, *Z. Phys.*, **37**, 263 (1926); M. Born, *Z. Phys.*, **40**, 167 (1927).

30. C. Davisson and C. H. Kunsmann, *Phys. Rev.*, **22**, 243 (1927); C. Davisson and G. H. Germer, *Nature (London)*, **119**, 558 (1927).

31. G. P. Thomson and A. Reid, *Nature (London)*, **119**, 890 (1927).

32. M. Born, *Z. Phys.*, **38**, 803 (1926).

33. N. Bohr, Atti del Congresso Internazionale di Como, Settembre 1927 (reprinted in *Nature (London)*, **121**, 78, 580 (1928).

34. W. Heisenberg, *Z. Phys.*, **43**, 172 (1927).

35. *Albert Einstein: Philosopher and Scientist*, The Library of Living Philosopher, Inc. Tudor Publishing Company, 1949, 1951, Harper & Row, New York, 1959. See also Ref. 4.

36. See, for example, B. D' Espagnat, *Conceptions de la Physique Contemporaine*, Herman, Paris, 1965, and the following articles in *Rev. Mod. Phys.*, **38**, (1966) by J. S. Bell (p. 447), and D. Bohm and J. Bub (pp. 453 and 470).

37. B. d'Espagnat, *Conceptual Foundations of Quantum Mechanics*, 2nd ed., Benjamin, New York, 1976. See also, by the same author, the article: "The Quantum Theory and Reality," *Sci. Am.*, **241**(5), 128, November (1979), and the critical remarks by V. Weisskopf followed by d'Espagnat answer: *Sci. Am.*, **242**, 8, May (1980).

38. This terminology is used by d'Espagnat who analyses the foundation of this world view in three premises or assumptions: (1) realism, that is, the doctrine that regularities in observed phenomena are caused by some physical reality whose existence is independent of human observation; (2) inductive inference is a valid mode of reasoning and can be applied freely so that legitimate conclusions can be drawn from consistent observations; (3) "Eistein separability," that is, no influence of any kind can propagate faster than the speed of light. Weisskopf, in his critical remarks (Ref. 37), warns about the use of ("misleading") expressions, which suggest that a renunciation to "local realistic views" may be equivalent to a renunciation of "realism" in general. The experiments considered here are in agreement with the predictions of quantum mechanics and therefore provide a clear support to the "realism" inherent to quantum mechanics, which is *different* from classical realism, in particular from "local realistic views."

39. A. Einstein, B. Podolski, and N. Rosen, *Phys. Rev.*, **47**, 777 (1935).

40. J. S. Bell, *Foundations of Quantum Mechanics*, Proceedings of the Enrico Fermi International Summer School, Course 40, Academic Press, New York, 1971.

41. *Structure et Propertés des Noyaux Atomiques*, Rapports et Discussions du Septiénne Conseil de Physique tenu à Bruxelles du 25 au 29 Octobre, Gauthier-Villars, Paris, 1934.

42. J. D. Cockcroft and E. T. S. Walton, *Proc. R. Soc. London, Ser. A*, **137**, 229 (1932).

43. M. L. E. Oliphant and E. Rutherford, *Proc. R. Soc. London, Ser. A*, **141**, 259 (1933).

44. E. O. Lawrence and M. S. Livingston, *Phys. Rev.*, **37**, 1707 (1931); **38**, 834 (1931); **40**, 19 (1932); **42**, 1950 (1932).

45. J. Chadwick, *Nature (London)*, **129**, 312 (1932); *Proc. R. Soc. London, Ser. A*, **136**, 692 (1932).

46. C. D. Anderson, *Phys. Rev.*, **43**, 491 (1933).

47. P. M. S. Blackett and G. P. S. Occhialini, *Proc. R. Soc. London, Ser. A*, **139**, 699 (1933).

48. L. Meitner and K. Philipp: *Naturwissenschaften*, **21**, 468 (1933); J. Curie and F. Joliot: *C. R. Acad. Sci. Paris*, **196**, 1105, 1581 (1933); C. D. Anderson and S. H. Neddermeyer, *Phys. Rev.*, **43**, 1034 (1933); P. M. S. Blackett, J. Chadwick, and G. Occhialini, *Nature (London)*, **131**, 473 (1933); *Proc. R. Soc. London, Ser. A*, **144**, 235 (1934).

49. B. Rossi, *Z. Phys.*, **82**, 151 (1933).

50. W. Heisenberg, *Z. Phys.*, **77**, 1 (1932); **78**, 156 (1932); **80**, 587 (1933).

51. E. Majorana, *Z. Phys.*, **82**, 137 (1933); *Ric. Scient.* **4**(1), 559 (1933).

52. E. Fermi: *Nuovo Cimento*, **11**, 1 (1934); *Z. Phys.*, **88**, 161 (1934).

53. F. Joliot and I. Curie, *C. R. Acad. Sci. Paris*, **198**, 254, 559 (1934); *J. Phys.*, **5**, 153 (1934).

54. E. Fermi, *Ric. Scient.*, **5**(1), 283 (1934); *Nature (London)*, **133**, 757 (1934); E. Fermi, E. Amaldi, O. D'Agostino, F. Rasetti, and E. Segrè, *Proc. R. Soc. London, Ser. A*,

146, 483 (1934); E. Amaldi, O. D'Agostino, E. Fermi, B. Pontecorvo, F. Rasetti, and E. Segrè, *Proc. R. Roy. Soc. London, Ser. A,* **146,** 522 (1935); E. Amaldi and E. Fermi, *Phys. Rev,,* **50,** 899 (1936).

55. *Les Particules Élémentaires,* Rapports et Discussions du huitième Conseil de Physique tenu à l'Université de Bruxelles du 27 Septembre ay 2 Octobre, 1948, R. Stoops, Brussels, 1950.

56. C. D. Anderson and S. H. Neddermeyer, *Phys. Rev.,* **51,** 884 (1937); **54,** 88 (1938).

57. B. Rossi, *Naturwissenschaften* **20,** 65 (1932).

58. H. Yukawa, Proc. Phys.-Math. Soc. Jpn., **17,** 48 (1935).

59. M. Conversi, E. Pancini, and O. Piccioni: *Nuovo Cimento,* **3,** 372 (1945); *Phys. Rev.,* **71,** 209 (1947).

60. C. M. G. Lattes, G. P. Occhialini, and C. F. Powell, *Nature (London),* **159,** 186 (1947).

61. D. Rochester and C. C. Butler, *Nature (London),* **160,** 855 (1947).

62. W. L. Gardner and C. M. G. Lattes, *Science,* **107,** 270 (1948); J. Burfening, E. Gardner, and C. M. G. Lattes, *Phys. Rev.,* **75,** 382 (1949).

63. L. de Broglie, *C. R. Acad. Sci. Paris,* **195,** 862 (1932); **197,** 536 (1932).

64. *La Théorie Quantique des Champs,* Rapports et Discussions du Douzième Conseil de Physique tenu à l'Université Libre de Bruxelles du 9 au 14 Octobre 1961, Wiley-Interscience, New York, and R. Stoops, Brussels, 1962.

65. A. Pais, *Phys. Rev.,* **86,** 513 (1952).

66. W. B. Fowler, R. P. Shutt, A. M. Thorndike, and W. L. Whittermore, *Phys. Rev.,* **93,** 861 (1954).

67. M. Gell-Mann and A. Pais, *Proc. Glasgow Conf. on Nuclear and Meson Phys.,* p. 342, Pergamon Press, London and New York, 1934; K. Nishijima, *Prog. Theor. Phys. (Kyoto),* **12,** 107 (1954).

68. This is true for energies below $\sim 10^3$ GeV, but at extremely high energies ($\geq 10^4$ GeV) cosmic rays will probably not be superseded by accelerators in any forseeable future.

69. O. Chamberlain, E. Segrè, C. Wiegand, and T. Ypsilantis, *Phys. Rev.,* **100,** 947 (1955).

70. M. Gell-Mann, Caltech Report CSTL-20 (1961), *Phys. Rev.,* **125,** 1067 (1962); Y. Ne'eman, *Nucl. Phys.,* **26,** 222 (1961); M. Gell-Mann and Y. Ne'eman, *The Eightfold Way,* Benjamin, New York, 1964.

71. This discussion of the hyperfine splitting of the hydrogen isotopes was still affected by some uncertainty originating in part from the inaccuracy in the value of the electrodynamic coupling constant α.

72. For the experimental tests on quantum electrodynamics at low energy, see, for example, E. Picasso, *Acta Leopoldina,* Suppl. 8 Bd. 44, p. 159 (1976), Deut. Akad. Naturforsh. Leopoldina; F. Combley and E. Picasso: p. 717 of *Proceedings 68th Course on Metrology and Fundamental Constants,* "International Summer School Enrico Fermi, Varenna, July 1976," Societe' Italiana di Fisica, Bologna, 1980.

73. Jade Collaboration: Desy 80/14, March 1980; Pluto Collaboration: Desy 80/01, January 1980.

74. T. D. Lee and C. N. Yang, *Phys. Rev.,* **104,** 254 (1956).

75. C. S. Wu, E. Amber, R. W. Hayward, D. D. Hoppes, and R. P. Hudson, *Phys. Rev.,* **105,** 1413 (1957); R. Garvin, L. Lederman, and M. Weinrich, *Phys. Rev.,* **105,** 1415 (1957); J. I. Friedman and V. L. Telegdi, *Phys. Rev.,* **105,** 1681 (1957); H. Frauenfelder, *Phys. Rev.,* **106,** 386 (1957).

76. B. Pontecorvo, *Phys. Rev.*, **72**, 246 (1947); O. Klein, *Nature (London)*, **161**, 897 (1948); G. Puppi, *Nuovo Cimento*, **5**, 587 (1948); T. D. Lee, M. Rosenbluth, and C. N. Yang, *Phys. Rev.*, **75**, 905 (1949); J. Tiomno and J. A. Wheeler, *Rev. Mod. Phys.*, **21**, 144 (1949).

77. B. Pontecorvo, *J. Exp. Theor. Phys. USSR*, **37**, 1751 (1951); A. Salam, *Proc. Seventh Annual Conf. on High Energy Physics 1957*, Interscience, New York, 1957; J. Schwinger, *Ann. Phys.*, **2**, 407 (1957); N. Cabibbo and R. Gatto, *Phys. Rev. Lett.*, **5**, 114 (1960).

78. S. S. Gerschtein and J. B. Zel'dovich, *JETP (USSR)*, **29**, 698 (1955), translation in *Sov. Phys. JETP*, **2**, 576 (1957); R. P. Feynman and M. Gell-Mann, *Phys. Rev.*, **109**, 193 (1958).

79. *Fundamental Problems in Elementary Particles Physics*, Proceedings of the Fourteenth Conference on Physics at the University of Brussels, October 1967, Interscience, New York, 1968.

80. T. Regge, *Nuovo Cimento*, **14**, 951 (1959).

81. G. F. Chew and S. Frantschi, *Phys. Rev. Lett.*, **7**, 394 (1961).

82. M. Gell-Mann, *Phys. Rev. Lett.*, **8**, 214 (1964); G. Zweig, CERN Preprint Th-492 (1964).

83. Y. Nambu, *Phys. Rev. Lett.*, **4**, 380 (1960); J. Bernstein, S. Fubini, M. Gell-Mann, and W. Thirring, *Nuovo Cimento*, **17**, 757 (1960).

84. *La conductibilité Électrique des Metaux*, Rapports et Discussions du Quatrieénne Conseil de Physique tenu a Bruxelles du 24 au 29 Avril 1924, Gauthier-Villars, Paris, 1927.

85. *Le Magnétisme*, Rapports et Discussions du Sixiénne Conseil de Physique tenu a Bruxelles du 20 au 25 October 1930, Gauthier-Villars, Paris, 1932.

86. J. C. Slater, *The Self-Consistent Field for Molecules and Solids, Quantum Theory of Molecules and Solids*, Vol. 4, Ch. 10, McGraw-Hill, New York, 1974.

87. *L'État Solide*, Rapports et Discussions du neuviénne Conseil de Physique tenu à l'Université Libre de Bruxelles du 25 au 29 September 1951, R. Stoops, Bruxelles, 1952.

88. F. C. Franck and W. T. Read, *Phys. Rev.*, **79**, 722 (1950); F. Seitz, *Imperfections in Nearly Perfect Crystals*, W. Shockley, ed., Ch. 1, Wiley, New York, 1952.

89. J. M. Edges and J. W. Michell, *Philos. Mag.*, **44**, 223 (1953); W. C. Dash, *J. Appl. Phys.*, **27**, 1153 (1956); S. A. Amelinckx, *Phys. Mag.*, **1**, 269 (1956).

90. *Les Éléctrons dans les Métaux*, Rapports et Discussions du dixiénne Conseil de Physique tenu a Bruxelles du 13 au 17 Septembre 1954, R. Stoops, Bruxelles, 1955.

91. J. Bardeen, L. N. Cooper, and J. R. Schrieffer, *Phys. Rev.*, **108**, 1175 (1957).

92. *Order and Fluctuations in Equilibrium and Non-Equilibrium Statistical Mechanics*, G. Nicolis, G. Dewel, and J. W. Turner, eds., Wiley, New York, 1981.

93. K. G. Wilson and M. E. Fischer, *Phys. Rev. Lett.*, **28**, 240 (1972); K. G. Wilson and J. Kogut, *Phys. Rev.*, **12C**, 75 (1974); K. G. Wilson, *Rev. Mod. Phys.*, **47**, 773 (1975).

94. L. P. Kadanoff, *Critical Phenomena*, Proceedings of the International School of Physics "Enrico Fermi," Course 51, M. S. Green, ed., Academic Press, New York, 1971.

95. G. Nicolis and I. Prigogine: *Self-Organization in Non-Equilibrium Systems*, Wiley-Interscience, New York, 1977.

96. E. L. Koschmieder, "Bénard Convection," *Adv. Chem. Phys.*, **26**, 177 (1974).

97. E. Arecchi, in: *Interaction of Radiation with Condensed Matter*, Vol. I, IAEA, Vienna (1977).

98. G. Nicolis and J. F. G. Auchmuty, *Proc. Natl. Acad. Sci. USA*, **71**, 2748 (1974); D. H. Sattinger, *J. Math. Phys.*, **19**, 1720 (1978).

99. *Symmetry Properties of Nuclei*, Proceedings of the Fifteenth Conference on Physics at the University of Brussells, 28 September to 3 October 1970, Gordon and Beach, New York, 1974.

100. E. P. Wigner, *Phys. Rev.*, **51**, 106, 447 (1937); E. P. Wigner and E. Feenberg, *Rep. Prog. Phys., Phys. Soc. London*, **8**, 274 (1941); P. Franzini and L. Radicati, *Phys. Rev. Lett.*, **6**, 322 (1963).

101. "La structure et l'Evolution de l'Univers," Rapports et Discussions de l'onziènne Conseil de Physique, tenu à l'Université de Bruxelles du 9 au 13 Juin 1958, R. Stoops, Bruxelles, 1958.

102. E. M. Burbidge, G. R. Burbidge, W. A. Fowler, and F. Hoyle, *Rev. Mod. Phys.*, **29**, 547 (1957); *Science*, **124**, 611 (1956).

103. A. A. Penzias and R. H. Wilson, *Astr. J.*, **142**, 419 (1965).

104. J. R. Oppenheimer and H. Snyder, *Phys. Rev.*, **55**, 455 (1939).

105. *The Structure and Evolution of Galaxies*, Proceedings of the Thirteenth Conference on Physics at the University of Brusselles, September 1964, Interscience, New York, 1965.

106. R. Giacconi, H. Gursky, F. Paolini, and B. Rossi, *Phys. Rev. Lett.*, **9**, 439 (1962).

107. S. Bayer, E. T. Byram, T. A. Chubb, and H. Friedman, *Science*, **146**, 912 (1964).

108. As announcement of the discovery of quasars one can take No. 4872 of *Nature (London)*, appeared on March 16, 1963 (Vol. 197), where a set of four articles touch on a few fundamental properties of two radio sources: 3C 273 and 3C 48. The second of the four articles, due to Martin Schmidt, is the core of the whole argumentation. The first article by C. Hazard, M. B. Mackey, and A. J. Shimmins (p. 1037) presents the "Investigation of the Radio Source 3C 273 by the method of Lunar Occultation." In the second paper, by Martin Schmidt (p. 1040) entitled "3C 273: A Star-Like Object with Large Redshift", the author shows that 6 lines (four of the Balmer series, one of Mg II, the other of O III) can be explained with a redshift of 0.158. The third paper by J. B. Oke (p. 1040), is entitled "Absolute energy distribution in the optical spectrum of 3C 273" and the fourth, by J. L. Greenstein and T. A. Matthews (p. 1041), "Redshift of the Unusual Radio Source 3C 48", for which they give a value of 0.3675 as weighted average of six relatively sharp lines.

109. The efficiency of man-made accelerators, defined as the output power in the beam devided by the total power supply is: Berkeley Bevatron, $\sim 3 \times 10^{-4}$; 28 GeV CERN-PS, $\sim 3 \times 10^{-4}$; Frascati 1 GeV electrosyncrotron, $\sim 1 \times 10^{-4}$.

110. *Astrophysics and Gravitation*, Proceedings of the Sixteenth Solvay Conference in Physics at the University of Brussels, 24–28 September 1973, Editions de l'Université de Bruxelles, Brussels, 1974.

111. R. W. Klebesadel, I. B. Strong, and R. A. Olson, *Astrophys. J.*, **182**, L85 (1973).

NOTE: On November 1982 a Solvay Conference devoted to "Higher Energy Physics: What are the possibilities for extending our understanding of elementary particles and their interactions to much greater energies?" was held in the University of Texas at Austin.

THE SOLVAY CONFERENCES
IN CHEMISTRY, 1922–1978

A. R. UBBELOHDE

Imperial College
London, England

On an occasion such as the present, three distinctive aims of the Solvay Conferences on Chemistry warrant special mention. First, as earlier speakers have said, they are international. Second, they are elitist in regard to the small number of invited papers on the chosen theme. Third, their themes have been chosen so far as possible because they have a riverhead character from which hopefully much further flow of chemical knowledge might be foreseen.

In 1912 when Ernest Solvay founded his Institutes there was nothing particularly novel about being international. Ever since Western civilization began, many centuries ago, leaders of knowledge wandering across frontiers have carried a ferment with them. Meetings among notable individuals from different countries were part of the informal development of chemistry at least from mid-eighteenth century. More formally, widespread aims to foster exchanges of knowledge between entire national academies culminated in 1897 with the first meeting of an International Association of Academics in Wiesbaden. At that meeting participants included the so-called cartel of German-speaking academies: Vienna, Göttingen, Leipzig, and Munich, together with the Royal Society of London, the Academie des Sciences of Paris, the Royal Academy of Sciences of Prussia, the Imperial Academy of St. Petersburg, and the National Academies of Washington and of Rome.* By 1914 this International Association had grown to a total of 24 member Academies from 16 countries, but it failed to survive disruptions from World War I. In any event, we may rightly point to the weakness of this kind of grandiose international exchanges, which lay in the very wide spectrum of sciences to be represented within each Academy, combined with the lack of financial resources adequate to support effective action.

---

* Compare P. Alter, *Notes and Records of the Royal Society,* **14,** 241 (1980).

The Solvay Foundations in 1912 had a more pragmatic as well as a more businesslike character. Their scope was planned under the scientific guidance of two committees of internationally eminent scientists and was deliberately much narrower than could be regarded as characteristic for any national academy. Financial resources were provided to support scientific discussions on a chosen theme among comparatively few invited experts. Ample arrangements were made for documentation, so that every invited speaker was helped as well as in some sense constrained to write a full monograph around his own work while in full progress. This gave a special flavour of actuality to what were at the same time authoritative review chapters in each Solvay Report on Chemistry.

It must not be forgotten that since the first Solvay Conference in Chemistry in 1922 human tasks of communication have become greatly enlarged, and now often involve much deeper and narrower specializations than even in 1922. While remembering these changes we may briefly look at Reports of past Solvay Conferences on Chemistry, to observe how far their pattern matched the intentions of their founder. Only the most prominent features can be mentioned in the space available.

As a sure sign of real leadership in chemistry, at least part of the work being done by any specialist must show evident foresight, as well as abundant actuality. "Talent spotting" for scientific innovators has in this sense always been a prime responsibility of the international Solvay committee of scientists organizing each conference program (see Table I). Foresight is one of the most important qualities of leadership, not least so in science. Happily, looking back, without claiming that opportunities were never missed, one can say from the record that past Solvay scientific committees have done their tasks well.

Indeed, the first chemistry conference in 1922 had a stroke of luck in the history of science, since it disclosed two riverheads from which much chemical knowledge has since flowed. In Dalton's Atomic Theory in 1808, combining weights of atoms had been regarded as unchangeable constants of nature. At the 1922 Solvay conference, a triad of monographs by Soddy, Aston, and Urbain overturned this belief completely. They showed that in any Daltonic element isotopes occurred as a mixture. Each isotope had a much more rational mass than the chemical atomic weight, and their mixture ratio was an accident of chemical evolution and could be modified by diverse physicochemical operations. We now think this ratio could in favorable cases give important clues about the cosmic origin of the chemical elements in evolution.

A second discovery even more far-reaching in its ultimate consequences was how to determine the molecular structure of matter by X-ray diffraction. This was also described in a 1922 Solvay Report by Sir

TABLE I.   Membership of the Scientific Committee for Chemistry

President	Members		
1922 Sir William Pope Cambridge	A. Job Paris Secretary	J. Perrin Paris Dony-Henault	
1925 Sir William Pope Cambridge	A. Job Paris F. Swarts Ghent Secretary	J. Perrin Paris F. M. Jaeger Groningen Dony-Henault	E. Briner Geneva J. Duclaux Paris
1928 Sir William Pope Cambridge	E. Briner Geneva A. Job Paris Secretary	J. Duclaux Paris J. Perrin Paris H. Wuyts	F. M. Jaeger Groningen F. Swarts Ghent
1931 Sir William Pope Cambridge	M. Bodenstein Berlin J. Duclaux Paris Secretary	E. Briner Geneva F. M. Jaeger Groningen H. Wuyts	M. Delepine Paris J. Perrin Paris
1934 Sir William Pope	J. Perrin Paris F. M. Jaeger Groningen M. Delepine Paris Secretary	E. Briner Geneva F. Swarts Ghent H. Wuyts	J. Duclaux Paris M. Bodenstein Berlin
1937 Sir William Pope	J. Perrin Paris F. M. Jaeger Groningen M. Delepine Paris Secretary	E. Briner Geneva F. Swarts Ghent H. Wuyts	J. Duclaux Paris M. Bodenstein Berlin
1947 Paul Karrer Zurich	H. Backer Groningen (C. Hinshelwood) Secretary	N. Bjerrum Copenhagen P. Pascal Paris J. Timmermans	(M. Delepine) Paris (R. Robinson)
1950 Paul Karrer Zurich	H. Backer Groningen (C. Hinshelwood) H. Wuyts Secretary	N. Bjerrum Copenhagen P. Pascal Paris J. Timmermans	Ch. Dufraisse Paris (R. Robinson)

TABLE I. *(Continued)*

President	Members		
1953 Paul Karrer Zurich	H. Backer Groningen	Ch. Dufraisse Paris	(C. Hinshelwood)
	K. Lindenstrom-Lang Copenhagen	P. Pascal Paris	(R. Robinson)
	H. Wuyts		
	Secretary J. Timmermans		
1956 Paul Karrer Zurich	H. Backer Groningen	Ch. Dufraisse Paris	(K. Lindenstrom-Lang)
	(P. Pascal) Paris	(R. Robinson)	A. R. Ubbelohde London
	Secretary J. Timmermans		
1959 A. R. Ubbelohde London	J. de Boer Limburg	G. Chaudron Paris	(Ch. Dufraisse)
	W. Kuhn Basel	K. Lindenstrom-Lang Copenhagen	(L. Pauling)
	A. Todd Cambridge	(G. Wittig)	
	Secretary R. Defay		
	Hon. Secretaries H. Wuyts, J. Timmermans		
1962 A. R. Ubbelohde London	J. de Boer Limburg	G. Chaudron Paris	(Ch. Dufraisse)
	W. Kuhn Basel	J. Mayer	(A. Nesmyanov)
	L. Pauling Cal. Tech.	(A. Tiselius)	(A. Todd)
	G. Wittig Heidelberg		
	Secretary R. Defay		
	Hon. Secretaries J. Timmermans, H. Wuyts		
1965 A. R. Ubbelohde London	G. Chaudron Paris	J. de Boer Limburg	(Ch. Dufraisse)
	J. E. Mayer California	(A. Nesmyanov)	(L. Pauling)
	(A. Tiselius)	(A. Todd)	(G. Wittig)
	Secretary R. Defay		
	Hon. Secretaries J. Timmermans, H. Wuyts		
1969 A. R. Ubbelohde London	G. Chaudron Paris	J. de Boer Limburg	J. E. Mayer California
	G. Wittig Heidelberg		
	Secretary R. Defay		
	Hon. Secretary J. Timmermans		

William Bragg. Some years later, as a young Dewar Research Fellow at the Royal Institution, I had many personal contacts with Sir William Bragg, who at that time was nearly 80. He once described to me a heated discussion between himself and Rutherford, to determine which had made scientific discoveries with the more far-reaching outcome. Bragg who was a modest man gave the palm to Rutherford's experiments on atomic energy, but Rutherford maintained that the experimental elucidation of the molecular structure of matter, in ways opened up by Bragg, would eventually have much wider influence for man's mastery over nature than his own nuclear energy researches. Even in 1980 it was disputable which was right. For our present purpose what was noteworthy was the way in which the work of each of them was spotted as a broad riverhead of chemistry at this earliest Solvay Conference.

By contrast, the conferences in 1926 and 1928 were much more conventional and rather an anticlimax. In fact, their chemical programmes were spread over too wide a range of chemical themes, which although interesting (the French phrase is "Actualités Scientifiques") had insufficient theoretical background for the discussions to probe them deeply and fruitfully. It should not be forgotten that quantum theories of chemical valence were only beginning to emerge in those years. The 1931 Conference on the constitution and configuration of organic molecules was rather better knit together as a whole and the 1934 Solvay Report on Reactions of Oxygen was a real success, narrow and deep. This Report still contains much matter worth taking further even now, a quarter of a century later.

Unfortunately, the following conference in 1937 displays the classic trap, which awaits all scientific committees, however eminent, who have the presentation of chemical innovation as a major aim. Though comprehensive the 1937 Report which deals with current knowledge about vitamins and hormones proved to be (and at that time could hardly hope to be) little more than a mere chemical catalogue of these novel substances. After a gap during World War II, the 7th Solvay Conference in 1947 picked up the promise of the 1922 Reports on isotopes in a brilliant way. It discussed how small but measurable differences between radioactive or nonradioactive isotopes were being used to investigate intricate problems in chemistry, with monographs by leaders such as F. Joliot, Bainbridge, Paneth, von Hevesy, and Calvin. The 1950 Conference on mechanisms of oxidation formed a valuable though less well integrated pendant to the 1934 Conference on Oxygen. Next, the innovative chemistry of proteins was discussed, fruitfully but rather untidily in 1953; this was later matched in 1959 by a conference on nucleoproteins. In 1956 the Solvay Conference had given its attention to some characteristic problems of Inorganic Chemistry, which marks its emergence in public opinion as a major and distinctive branch of chemistry.

Subsequent Solvay Conferences have followed similar patterns. 1962 ushered in one of the best as well as most comprehensive reports on chemical kinetic theory, dealing with an exceptional range of transfer processes between molecules in gases. Since then other Solvay Conferences in Chemistry have been quite as comprehensive, but on the whole less well integrated into their chosen theme. No evident decline in quality can be perceived.

Even so, since 1922 there have been major transformations in the chemical sciences, as well as in the number of scientists working in them. The number of scientific conferences held each year has grown enormously throughout the world. This wealth of meetings, some of rather ephemeral nature, makes it harder to plan elitist discussions 2 or 3 years ahead, if one attempts to adopt the Solvay pattern to ensure really authoritative Reports. World competition between conferences of every quality also makes it much harder to bring even five or six elite scientists together in any chemical speciality, especially when the aim is to secure live monographs of the highest quality. Furthermore, the enormous growth of chemical knowledge imposes increasing specialization almost as a condition for outstanding excellence. This tends to hamper a uniform level of elitism.

All these considerations illuminate the foresight of Ernest Solvay when he envisaged liquidation after a few decades of his very successful foundation of Chemical Institutes in their original form. There are, of course, other patterns; the evolution of international conferences in Chemistry proceeds without interruption. The present is an auspicious occasion. We are about to experience a Solvay pattern in another way, at this first joint conference of the Solvay Institutes with the U.S. National Academy of Sciences. A few carefully selected specialists have been brought together to discuss a quite limited range of topics around a central theme. From past experience of Solvay Conferences, this is a perennial means to achieve success. We may indeed look forward to stirring and fruitful discussions.

NONEQUILIBRIUM THERMODYNAMICS AND CHEMICAL EVOLUTION: AN OVERVIEW

I. PRIGOGINE

Faculté des Sciences de l'Université Libre de Bruxelle
and
Center for Studies in Statistical Mechanics
University of Texas at Austin

I. INTRODUCTION

Let me first explain why the Solvay Institutes have chosen the subject "Aspects Chemical Evolution" for this very special occasion. This is, of course, not a usual Solvay Conference. It takes place, as has been mentioned, on the occasion of the 150th anniversary of the independence of Belgium. Belgium being a small country, we had necessarily to concentrate on fields in which its contribution has been considered as significant by the scientific community. The problem of chemical evolution satisfies this criterion as it combines aspects of importance from the thermodynamic point of view with atmospheric and biological research. These are fields in which, traditionally, there has been a great deal of activity in Belgium.

Before addressing the main topic of chemical evolution, I would like to discuss briefly the rather curious story of the Belgian school of thermodynamics, often called the "Brussels school." It took shape at the end of the 1920s and during the 1930s. At a time when the great schools of thermodynamics, such as the Californian school founded by Lewis and the British school with Guggenheim, directed their efforts almost exclusively to the study of *equilibrium systems,* the point of view presented by the Brussels school appeared as quite unorthodox and somewhat controversial. Indeed, the Brussels school tried to approach equilibrium as a special case of nonequilibrium and concentrated its efforts on the presentation of thermodynamics in a form that would be applicable also to nonequilibrium situations. This story is rather curious from the point of view of the history of science, so let me go into a little more detail.

43

As you may know, Belgium was the first industrialized country on the European continent, following very rapidly the British industrialization. The first heat engine was constructed in Belgium in 1721, and about a century later, the first railroad on the continent ran between Brussels and Malines. Not unexpectedly, there was a great interest in technical thermodynamics during all of the nineteenth century and the beginning of the twentieth. Let me only mention that the early research on diesel engines was done by Diesel, a German citizen, in Ghent, a Belgian town.

Now I have to switch to my teacher, De Donder, born in 1872. DeDonder was a self-taught scientist. He was too poor to study at a university, but it was then possible in Belgium to obtain a degree, without going to the university, by passing state examinations. He became a school teacher and, while performing his duties at the school, contributed significantly to the general theory of relativity. He was in close correspondence with Einstein and Lorentz, and we find many references to his work in Eddington's presentation of relativity.[1,2] In 1918, he became a professor at the Faculty of Applied Sciences of the University of Brussels. As he told me personally, he had no engineering background, no special knowledge of technical problems, but was convinced that engineers should know thermodynamics. For this reason he began to prepare a set of lectures on thermodynamics for engineers.

At that time he knew only two books on thermodynamics: "Theory of Heat" by Clausius and the classical works of Gibbs[3] which were translated into French by Le Chatelier. In these two books he found what looked to be two conflicting views on thermodynamics. As is well known, Clausius wrote the second law in the form:

$$dS = \frac{dQ}{T} + \frac{dQ'}{T} \tag{1}$$

where dQ is the heat received by the system during a time interval, say dt, T the temperature, and dQ' the so-called "uncompensated heat," which may correspond, for example, to losses in real thermal engines. The second law as expressed by Clausius was that the uncompensated heat could only be positive or vanishing: $dQ' \geq 0$. Clausius' formula is written today more generally as

$$dS = d_e S + d_i S \tag{2}$$

where $d_e S$ corresponds to entropy flow and $d_i S$, to the entropy production (Fig. 1).

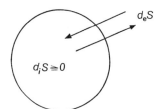

Fig. 1.   Open systems: Internal entropy production, d_iS ≥ 0. d_eS is the exchange of entropy with the environment.

In contrast to Clausius, Gibbs did not discuss uncompensated heat, as he started directly with the *total differential of entropy*. Gibbs' presentation appealed very much to De Donder. However, he wanted to find the meaning of this mysterious "uncompensated heat." He considered a system whose physical conditions, such as pressure and temperature, were uniform and which was closed to the flow of matter. Chemical reactions, however, could go on inside the system. De Donder first introduced what he called the degree of advancement, ξ, of the chemical reaction so that the reaction rate v is the time derivative of ξ.

$$v = \frac{d\xi}{dt} \tag{3}$$

Now, it is easy to see that under these conditions the uncompensated heat can be written as the differential form

$$dQ' = A_{p\xi}\, dT + A_{T\xi}\, dp + A_{Tp}\, d\xi \geq 0 \tag{4}$$

where $A_{p\xi}$, $A_{T\xi}$, and A_{Tp} are suitable coefficients. De Donder then introduced the assumption that the chemical reactions are the only irreversible processes in the system. This requires that

$$A_{p\xi} = A_{T\xi} = 0 \quad \text{and} \quad dQ' = A_{Tp}\, d\xi \tag{5}$$

As a result, the uncompensated heat must be the same whatever the set of physical variables, pressure and temperature or volume and temperature, is used to describe the state of the system. Indeed, we have

$$A_{Tp} = A_{TV} = \cdots = A \tag{6}$$

This means that A is independent of the external constraints acting on the system. It must be a *state function* like energy or entropy. Moreover, in terms of the entropy production per unit time, the Clausius inequality now becomes

$$\frac{d_iS}{dt} = \frac{A\mathbf{v}}{T} \geq 0 \tag{7}$$

We may now see why De Donder called this new state function the "affinity" of the chemical reaction. It has always the same sign as the reaction rate and vanishes at equilibrium. This is somewhat similar to the temperature gradient which determines the direction of the flow of heat.

The introduction of affinity by De Donder marks the birth of the Brussels school; the first publication appeared around 1922, but it took some years to make these concepts more precise.[4] What was the reaction of the scientific community? When we go through the proceedings of the Belgian Royal Academy, we see that De Donder's work indeed aroused much local interest. Verschaffel from Ghent and Mund from Louvain were among the people who became active in this newborn nonequilibrium chemical thermodynamics. However, one has to say that elsewhere De Donder's approach met with skepticism and even with hostility. His introduction of affinity was thought of as merely a different notation.

Of course, De Donder was not the only thermodynamicist who was interested in nonequilibrium conditions at this time. The work of the French school, notably Duhem and Jouguet, and the Cracow school with Natanson should also be recalled. (A short survey of the history can be found in my *Etude Thermodynamique des Phĕnomènes Irreversibles,* Desoer, Liege, 1947.) However, his contributions to the introduction of the degree of advancement and of affinity are now classical.

It is true that in his own work De Donder did not pursue the consequences of nonequilibrium very far. We have to wait until the basic discovery of Onsager's reciprocity relations in 1931 and till the work of Eckart, Meixner, and many others in the 1940s and the 1950s to see thermodynamics of nonequilibrium processes take shape and be integrated into common knowledge.

II. BIFURCATIONS IN NONEQUILIBRIUM SYSTEMS: DISSIPATIVE STRUCTURES

Today the situation has changed. We understand now that the laws of thermodynamics are universal at and near equilibrium, but become highly specific in far-from-equilibrium conditions. We also realize that irreversible processes can become a source of order. We see irreversible processes taking a prominent role in our description and understanding of nature. If the proton, as is thought to be by many physicists, turns out to be an unstable particle, then all nuclei are unstable and will ultimately decay; the structure of matter as we know it today is not eternal but is in the process of irreversible disintegration. The primordial 3°K blackbody radiation, predicted in the 1940s[5] and experimentally discovered in the 1960s, persuades us to think of an evolutionary processes at the cosmological level. An evolutionary process in which, of the many possible histories, only one becomes a reality: a sequence of events resulting in a singular fate, as for instance, a universe in which there is an excess of baryons over antibaryons. In ordinary macroscopic systems, we see a similar phenomenon arising due to far-from-equilibrium conditions. Chemical evolution is a consequence of the irreversible processes.

Now irreversible processes can be studied on three levels: (1) the phenomenological thermodynamic level in which the equations for the macroscopic variables are studied; (2) the level of fluctuations, in which we study the nature, growth or decay of small fluctuations either of internal or external origin; and (3) the "basic" level in which we try to identify the microscopic mechanisms of irreversibility. Here I shall be mainly concerned with the first, and to certain extent the second level, through which I believe we can begin to understand how irreversible processes bring about the different aspects of the process of evolution.

At the phenomenological thermodynamic level, when we go far from equilibrium, the striking new feature is that new dynamical states of matter arise. We may call these states dissipative structures as they present both structure and coherence and their maintenance requires dissipation of energy.[6] Dissipative processes that destroy structure at and near equilibrium may *create* these structures when sufficiently far from equilibrium.

Irreversible processes correspond to the time evolution in which the past and the future play different roles. In processes such as heat conduction, diffusion, and chemical reaction there is an arrow of time. As we have seen, the second law postulates the existence of entropy S, whose time change can be written as a sum of two parts: One is the flow of entropy $d_e S$ and the other is the entropy production $d_i S$, what Clausius called "uncompensated heat,"

$$dS = d_i S + d_i S$$

As we stated earlier, $d_i S \geq 0$. Generalizing the concept of affinity of chemical reactions, the rate of entropy production can be written as[7]

$$\frac{d_i S}{dt} = \sum_\alpha J_\alpha X_\alpha \tag{8}$$

in which the J_α's are the flows or rates of irreversible processes, and X_α's are the corresponding forces such as gradients of temperature or of concentration. Much of modern thermodynamics is centered around this equation and the relation between the forces X_α and the flows J_α.

We can now distinguish three stages in the development of thermodynamics. First, there is the equilibrium stage in which the forces and the consequent flows vanish. It is under those conditions that we have equilibrium phase transitions such as solid to liquid and liquid to vapor. The structures that arise in such phenomenon, as for instance in a crystal, can be understood in terms of the minimization of the well-known free energy F. We have

$$F = E - TS \tag{9}$$

where E is the total energy, T the temperature, and S the entropy. Equilibrium corresponds to competition between energy minimization and entropy maximization at a given temperature. Next, we have the near-equilibrium regime in which flows are proportional to forces. In this *linear* regime, the celebrated Onsager relations are valid. In these two regions the stability of the state is guaranteed; small fluctuations are damped by the dissipative processes. We have a basically homeostatic description of the physical world. In the third, far-from-equilibrium stage, this is no longer so. Here the fluctuations may be amplified and change the macroscopic pattern of the system. The state of the system extrapolated from the equilibrium state becomes unstable and, through fluctuations, is driven into a new state. Instability means that under these conditions existing structure cannot be maintained; the system has now become open to new possibilities. As we shall see in more detail soon, the system becomes open to a multiplicity of new possibilities and the phenomenological equations above cannot predict which of the new possibilities will be realized. Evolution of the nonequilibrium structure depends on some singular factor that happens to occur at the right time, at the right place: it depends on chance.

We shall now illustrate these aspects through the far-from-equilibrium reaction–diffusion systems, which have been studied in great detail in recent years and which may be taken as a kind of archetype in this context. If chemical reaction and diffusion are the only relevant dissipative processes to be considered, the equations that describe the spatiotemporal variation of the concentrations $\mathbf{X} \equiv (X_1, X_2, \ldots, X_N)$ are

$$\frac{\partial}{\partial t} \mathbf{X} = \mathbf{v}(\mathbf{X}, \{\lambda_i\}) - \operatorname{div} \mathbf{J_x} \tag{10}$$

Here $\mathbf{v}$ is the rate of production (or consumption) of the reactants, $\{\lambda_i\}$ a set of control parameters that keep the system far from equilibrium, and $\mathbf{J}_x$ is the diffusion flow which is usually considered Fickian, that is, $\mathbf{J}_i = -D_i\,\mathrm{grad}\,X_i$. In the vicinity of equilibrium there is a stationary solution $\mathbf{X}_s$, that is, $\partial\mathbf{X}_s/\partial t = 0$. This solution $\mathbf{X}_s$, of course, depends on the set of parameters $\{\lambda_i\}$ and it changes as these parameters are varied. For a given set of values of the parameter, the stability of the state $\mathbf{X}_s$ is investigated by defining $\mathbf{x}$ so that $\mathbf{X} \equiv \mathbf{X}_s + \mathbf{x}$, and looking at the evolution of the small deviation $\mathbf{x}$. If the system is stable (more precisely, asymptotically stable), $\mathbf{x}$ will tend to zero as is the case near equilibrium. However, under far-from-equilibrium conditions, which are obtained here through the variation of $\{\lambda_i\}$, there could arise an instability and the appearance of new stable solutions to equation (10) which are not extrapolation of $\mathbf{X}_s$. Mathematically this is the phenomenon of bifurcation of new solutions. The value of $\{\lambda_i\}$ at which the solution $\mathbf{X}_s$ becomes unstable and at which new solutions appear is called the bifurcation point. If the new solutions are also steady states, then we have bifurcation of steady states. When one does the mathematical analysis of bifurcation, one finds that the new solutions are characterized by their "amplitudes" $\{\alpha_i\}$, which are solutions of a set of nonlinear equations called the bifurcation equations. In a simple case, there is one parameter λ, and two new solutions bifurcating at the bifurcation point λ_c. This is schematically represented by a bifurcation diagram shown in Fig. 2.

The steady-state solution that is an extension of the equilibrium state, called the "thermodynamic branch," is stable until the parameter λ reaches the critical value λ_c. For values larger than λ_c, there appear two "new branches" ($b1$) and ($b2$). Each of the new branches is stable, but the extrapolation of the thermodynamic branch (a') is unstable. Using the mathematical methods of bifurcation theory, one can determine the point λ_c and also obtain the new solution, (i.e., the dissipative structures) in the vicinity of λ_c, as a function of ($\lambda - \lambda_c$). One must emphasize that

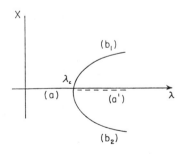

Fig. 2.   A typical bifurcation diagram. λ_c is the bifurcation point at which the branch (a) becomes the unstable branch (a'), and two new stable branches ($b1$) and ($b2$) bifurcate.

the phenomenon of bifurcation is extremely general and that there is a wide variety of solutions that can bifurcate: multiple steady states, oscillating solutions, propagating waves, etc. Also there could occur higher-order bifurcations: for values of λ larger than λ_c, the branches $b1$ and $b2$ could themselves become unstable leading to higher order structures. Most of the general features of evolution of order in nonequilibrium systems, however, are already present in the simple case represented in Fig. 2. If we were considering a one dimensional system, for instance, the two branches ($b1$ and $b2$) could correspond to the two possible structures shown in Fig. 3 in which we have represented the concentration of one of the reactants as a function of position. Here we see that we either have a "left" or a "right" directed structure. The choice between the two possibilities is not in equation (10); just from this equation it is impossible to decide which of the possibilities will be realized. The selection of one of the two structures will depend on some external factor or perhaps some unique event that will leave a permanent imprint on the system by the selection. It is this feature that will make the formation of structures *sensitive* to the environment. We will discuss this point in greater detail later. Here we note that nonequilibrium conditions lead to multiple possibilities out of which only one is realized.

In the example given above, in a system where initially there was no preferred direction, after the bifurcation there is now a sense of direction. In this way, the first bifurcation introduces a single space or time parameter. But this is only the start. As mentioned earlier, there are secondary and higher bifurcations that are possible. Figure 4 shows a schematic diagram of such successive bifurcations. The successive bifurcating solutions might be quite different in character.

One of the well-studied systems that illustrates this successive-bifurcation behavior is the Belousov–Zhabotinski reaction. Let me briefly show you the results of some experiments done at the University of Texas at Austin,[8] referring for further details to the discussion by J. S. Turner in this volume. The experimental setup of the continuously stirred reactor

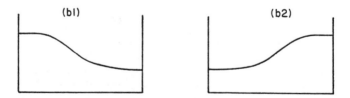

Fig. 3.   An example of dissipative structures corresponding to the two branches ($b1$) and ($b2$) of Fig. 2. Here concentration of one of the reactants is shown as a function of position.

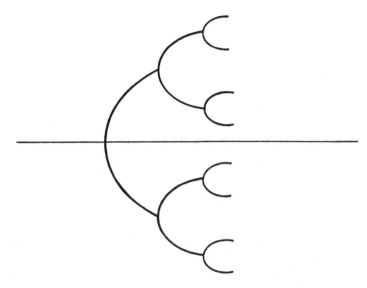

Fig. 4.   Schematic representation of successive bifurcations leading to complex structures.

through which there is a flow of reactants is shown in Fig. 5. Close to equilibrium the system is in a steady state. As the system is driven away from equilibrium it goes through an instability and begins to oscillate. These oscillations, at first nearly sinusoidal, become more and more complex through successive bifurcation. Figures 6, 7, and 8 show the temporal variation of the concentration of Br^- ion as a function of time along with power spectrum that tells us the Fourier frequency components of the oscillations. It can be seen that with increasing distance from equilibrium

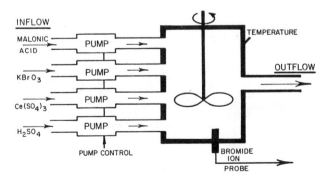

Fig. 5.   Experimental arrangement of the continuously stirred Belousov–Zhabotinski reaction.

the oscillations become more and more complex and approach "chaotic" behavior. This chaos is very distinct from the thermal chaos: it is a macroscopic chaos with macroscopic wavelengths and frequencies for the chaotic behavior of macroscopic variables. Such behavior arising out of relatively simple nonlinear equations has been the subject of considerable interest recently. One of the somewhat surprising features is the discovery of the "strange attractor." In systems of ordinary differential equations, usually, an attractor to which all neighboring trajectories converge is either a point or a trajectory. But in a large class of nonlinear systems it is found that the attractor, called the strange attractor, is not a topological

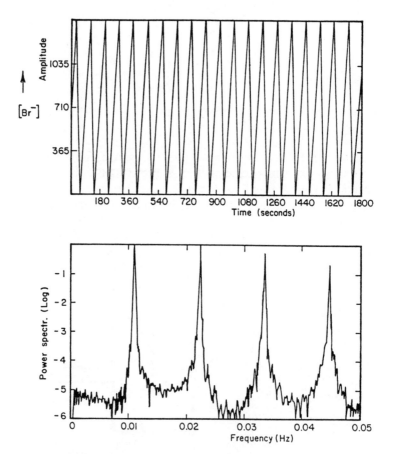

Fig. 6.   Regular oscillations and the corresponding power spectrum of the Belousov–Zhabotinski reaction.

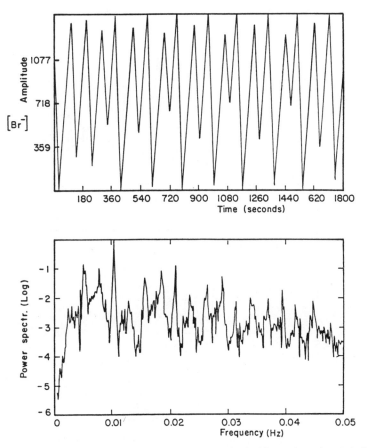

Fig. 7.   Mixed mode oscillations in the Belousov–Zhabotinski reaction when it is farther from equilibrium than it is in Fig. 6.

manifold; it is due to this fact that some systems exhibit a chaotic behavior. In this connection I would like to mention that for discrete maps also chaotic behavior can occur. Recently, Mitchel Feigenbaum discovered that in a large class of discrete maps the approach to chaotic behavior through successive bifurcation has both qualitative and quantitative universality. There are means by which the behavior of continuous systems can be reduced to that of a discrete map for which this universality may hold, but it is not clear what is lost in the process.

It is remarkable that farther away from this chaotic region, the Belousov–Zhabotinski system returns to more orderly behavior. All the be-

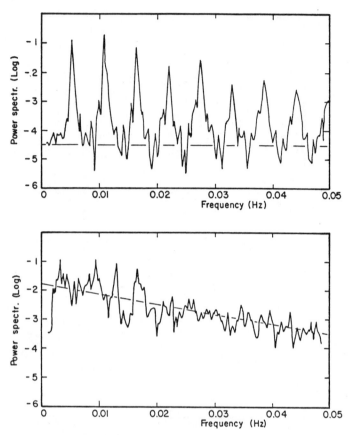

Fig. 8. When the Belousov–Zhabotinski reaction is sufficiently far from equilibrium it shows a chaotic behavior. This is reflected in the power spectrum being flat in comparison with the spectrum of the more orderly oscillatory behavior.

havior of this system is schematically summarized in Fig. 9. We may contrast this with the "peaceful" description of equilibrium diagrams.

III.   SENSITIVITY AND SELECTION

So far, we have seen how irreversible processes can become a source of order. We have seen how, through successive bifurcations, the system can become more and more complex. This has *some* features of the process of evolution, but does not contain many of the essential features. First of all, there is a certain overall irreversible feature in the process of evolution. Structures and organization appear in a sequential way. In the system we considered in the previous section, this is not so. The parameters of this system are controlled externally and depending on the manner in which they are varied the sequence of structures can appear in one direction or its opposite. In the process of evolution we do not see this invertibility. One of the mechanisms through which this can happen is when the nonequilibrium conditions (the parameters) themselves are evolving irreversibly so that a structure appears and is "frozen" due to changing conditions. The spontaneous appearance of certain sequences of RNA, depending on the environmental conditions, has been investigated and experimentally demonstrated by Manfred Eigen and is discussed in this contribution in the present volume. The particular class of sequences that appeared in the early prebiotic evolution, due to mechanisms that are much more complex than the one mentioned above, are in a sense "frozen" in that state. If the appearance of a certain sequence involves overcoming a barrier, then the peculiar conditions at the appropriate time will create a particular sequence that will then dominate and become permanent. The processes of freezing of nonequilibrium structures can be seen most clearly in the zygote of the seaweed *Fucus*. The membrane of the spherical zygote has active pumps and passive channels

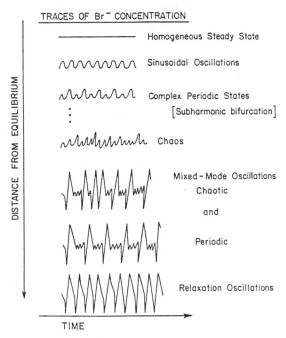

Fig. 9.   A schematic representation of the different types of nonequilibrium behavior in the Belousov–Zhabotinski reaction.

for small ions; these can freely diffuse on this membrane. Due to nonequilibrium conditions they aggregate at two diametrically opposite points
and create an axis that plays an important role in the development of the
Fucus.[9,10,11] It is known that, after the formation of this axis, the metabolic
conditions of the system change and the pumps are "frozen" in this configuration: though the nonequilibrium conditions that created this structure are no longer present, the structure, which has important consequences, is left behind.

One can see such frozen structures that were formed due to an autocatalytic reaction in tektites.[12] There are several other geological processes that show such "freezing" of nonequilibrium structures.[13]

In addition, there is another interesting nonequilibrium mechanism that
can produce one type of structure which then remains permanently. Suppose there was a far-from-equilibrium chemical system with three reactants X, Y, and Z that oscillate. As in the case of the Belousov–Zhabotinski reaction, let us assume that the concentrations of these variables
reach their maxima in a well-defined order: X reaches its maximum first
followed by Y and Z successively. The order $X \rightarrow Y \rightarrow Z$ is determined
(and fixed) by the nonequilibrium kinetics. Now suppose that such a system is coupled to a polymerizing catalyst that can produce either of the
following two unidentical polymers:

$$X\text{-}Y\text{-}Z\text{-}X\text{-}Y\text{-}Z\text{-}X\text{-}Y\text{-}Z \quad \text{(A)}$$

$$X\text{-}Z\text{-}Y\text{-}X\text{-}Z\text{-}Y\text{-}X\text{-}Z\text{-}Y \quad \text{(B)}$$

(Note that this implies that each monomer X, Y, Z has a sense of direction;
if not, one is obtained by rotating the other through 180°.) In such a system,
since the abundance of X, Y, and Z occurs in a well-defined order, polymer (A) will be synthesized in much greater quantities than polymer (B).
After some time, if the oscillations die out due to some reason, they will
have left behind an abundance of the polymer (A). One important feature
of this example is the link between macroscopic symmetry breaking and
microscopic symmetry breaking. It points to the interesting possibility of
macroscopic symmetry breaking states being the cause of microscopic
asymmetry.

The next feature that I would like to discuss is the notion of selection
that is central to the concept of evolution. That selection occurs even at
the level of prebiotic conditions has been well demonstrated by Eigen
and his group. In these experiments it was shown that several RNA sequences compete and of the many possibilities only a small fraction of

similar sequences are selected by the environmental conditions. In this context I would like to point out that nonequilibrium systems become *sensitive* to small external factors and this aspect is important in understanding selection at a basic thermodynamic level. That nonequilibrium systems develop sensitivity that equilibrium systems do not have is clear when we consider a phenomenon like Bénard convection: The convection pattern that forms is a response to the gravitational field. In contrast to the extremely small effect of a gravitational field on a fluid in equilibrium, under the nonequilibrium condition of sufficiently large temperature gradient, gravity has the large effect of causing a convection pattern. We have recently analyzed the effects of an external field in the *selection* of a nonequilibrium structure[14] and found that, due to the nonequilibrium sensitivity, this selection process is extremely efficient. The system we considered is a nonequilibrium chemical system discussed earlier. In the absence of the external field, when the system is beyond the critical point, two patterns, as shown in Fig. 3, are possible with the corresponding bifurcation diagram shown in Fig. 2. As there is no intrinsic direction in the system, both structures are equally likely (this is also reflected in the symmetry of the bifurcation diagram). When we introduce the field, the symmetry of the system is broken as we have introduced a direction. Bifurcation analysis shows that the corresponding bifurcation diagram is as in Fig. 10. Here, the two branches are separated, implying that the system will evolve to one of the structures, say ($b1$) as in Fig. 10, and not to the other, unless there is a large enough fluctuation. Thus, the external field selects one of the two possible structures. The effectiveness with which this selection occurs depends on the separation s between the

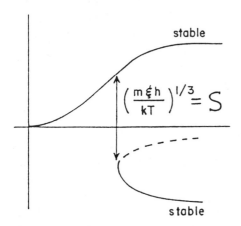

stable

$$\left(\frac{m\xi h}{kT}\right)^{1/3} = S$$

stable

Fig. 10.   Sensitivity of nonequilibrium systems: The bifurcation diagram in the presence of an external field g. s is a measure of the sensitivity of the system to the field.

two branches. Detailed calculation shows that

$$s = k_0 \left(\frac{\xi g L}{k_B T}\right)^{1/3}$$

(11)

in which ξ is coupling constant for the interaction of the molecules with
the field, g is the strength of the field, L the size of the system, k_B the
Boltzmann constant, and T the temperature. k_0 is a factor that depends
on the chemical kinetics. If we consider an equilibrium system and look
at the response such as the formation of a concentration gradient we would
find that it is proportional to $(\xi g L / k_B T)$, without the fractional exponent.
For most equilibrium systems $(\xi g L / k_B T) \ll 1$, as for instance when g is the
gravitational field for which ξ can be taken to be the atomic mass unit.
For nonequilibrium systems, however, due to the fractional exponent,
the effect is *greatly* enhanced. Detailed calculation for a specific model
were performed[14] to verify the quantitative aspects of such high sensitivity. Similar analysis was also done for rotating waves in the presence
of circularly polarized light.[15] The nonequilibrium sensitivity we discussed here is a very general phenomenon as it depends on bifurcation
and symmetry breaking which are very general. In the context of selection
of biomolecules, this should enable us to understand the role of nonequilibrium conditions in the process of selection more quantitatively.

IV.   STOCHASTIC ASPECTS OF EVOLUTION

As fluctuations are an intrinsic part of a thermodynamic system, a
discussion of nonequilibrium structures is not complete without the consideration of the consequences of fluctuations. Unlike equilibrium systems, nonequilibrium systems do not have a general prescription, like the
Einstein formula, to describe the fluctuations. Nonequilibrium fluctuations are highly specific. The importance of fluctuations appears clearly
in the way they alter the macroscopic behavior in the vicinity of the
bifurcation point and also in the way the coherence of a structure depends
on the dimensionality of the system in the face of the destructive influence
of fluctuations.

Fluctuations in nonequilibrium systems have been studied mainly
through two approaches: the master equation approach[16] and more recently the Ginsburg–Landau functional approach.[17] In the master equation approach, the microscopic transition probabilities for chemical reactions and diffusion are taken to be given, and a master equation for the
spatiotemporal variation of the probability distribution is obtained.
Though the explicit solution of the master equation is difficult to obtain,
some important general features could be deduced from it. One can show

that chemical kinetics in the nonequilibrium regime generates non-Poissonian distribution. When diffusion is also considered, it can be seen that it tends to make the distribution locally Poissonian counteracting the nonequilibrium kinetics. One also finds that whenever there is a deviation from the Poissonian nature there is an accompanying spatial correlation. This correlation length is the characteristic length a molecule will diffuse before it is transformed by chemical reaction. Several such interesting nonequilibrium aspects of fluctuations can be deduced through this approach.

The second approach is more intuitive and depends on separation of processes of different time scales. The "fast" processes are taken to be the fluctuations that accompany the "slow" processes and a Langevin-type equation is written for the slow variables: the slow variables describe the macroscopic nonequilibrium structures. From such considerations a functional, called the Ginsburg–Landau functional, that gives the probability distribution can be obtained. This approach has been very successfully used for equilibrium systems, because one can obtain the appropriate noise term using the fluctuation–dissipation theorem. Using the modern renormalization group methods, several aspects of the influence of fluctuations on critical behavior can be derived. For instance, one can obtain a relation between the stability of a structure and dimensionality: if the dimension d of the system is less than a certain critical dimension d_c, the structure will be destroyed by fluctuations. For nonequilibrium systems, below and near the first bifurcation, this method has been applied quite successfully.[17] The application of this method to nonequilibrium symmetry breaking transitions has been questioned.[18] This is because in some symmetry breaking nonequilibrium transitions, like the Bénard instability, the characteristic lengths are not intrinsic, but are dependent on the boundary conditions. For the diffusion–reaction systems however, it is now clear that there is such an intrinsic length and that the renormalization group methods could be applied. One must note that this method faces the difficulty of not having a well-defined way to get the appropriate fluctuating term in the far-from-equilibrium region where there are multiple steady states, etc. The master equation approach does not have such ambiguities and most of the intuitive aspects of the Ginsburg–Landau approach can be obtained from it.[16]

A knowledge of the magnitude and the nature of fluctuations is essential for an understanding of the nonequilibrium sensitivity. The influence of an external factor will be significant only when it can overcome the randomizing influence of the fluctuations; both sensitivity, obtained from macroscopic equations, and a knowledge of fluctuations are required for a proper understanding of the process of selection.

In situations where the macroscopic equations tell us that a coherent state is possible, it is important to know if this coherence could withstand the destructive influence of fluctuations. For instance, the macroscopic equations of a diffusion-reaction system might have a solution that implies that the concentrations can oscillate homogeneously. We expect the inhomogenous fluctuations to have a tendency to destroy this coherence if they are "strong" enough. An analysis of the situation shows that the coherence can be maintained if the dimension of the system is larger than a certain critical dimension. Malek-Mansour et al.[19] have shown that for chemical systems amenable to one variable, the critical dimension d_c is given by

$$d_c = 2 \frac{k + 1}{k - 1} \tag{12}$$

where k is the degree of the leading nonlinearity of the chemical reaction. If $d_c > 3$ then the coherence will tend to be destroyed by inhomogeneous fluctuations. Thus we see that we require $k \geq 5$ for coherence to be secured in three dimensions and coherence will tend to be compromised in systems with dimension less than three. Similar conclusions can be arrived at through the Ginsburg–Landau approach and renormalization group methods.

Order arising through nucleation occurs both in equilibrium and non-equilibrium systems. In such a process the order that appears is not always the most stable one; there are often competing processes that will lead to different structures, and the structure that appears is the one that nucleates first. For instance, in the analysis of the different possible structures in diffusion–reaction systems[17,20] one can show, by analyzing the bifurcation equations, that there are several possible structures and some of them require a finite amplitude to become stable; if this finite amplitude is realized through fluctuation, this structure will appear. In the formation of crystals (hydrates) the situation is similar: the structure that is formed depends, according to the Ostwald rule, on the kinetics of nucleation and not on the relative stability.

V. CONCLUDING REMARKS

In summary, we see that the study of nonequilibrium systems gives us some understanding of the process of evolution. The appearance of a certain type of organization or order is through fluctuations and kinematic considerations. In general we cannot associate a strict optimization principle to evolution: there is no strict "Darwinian" selection of *the* fittest.

In the discussion of fluctuations above, the mechanisms that are involved are given and the structure that arises is a consequence of these mechanisms. But evolution can occur through the appearance of new mechanisms. The interaction between the new mechanism and the existing structure will then determine the fate of both the mechanism and the structure. Such aspects need more study.

Nonequilibrium macroscopic conditions create new microscopic processes which in turn change the macroscopic conditions and so on. On a large scale we can see such a feedback mechanism at work during the early stages of evolution. The atmosphere then did not contain as much oxygen as it now contains. The early forms of life deposited oxygen in the atmosphere and in course of time the oxygen content increased and new forms of oxygen-dependent life could evolve. These forms of life in turn are changing the environmental conditions: Man is changing the environment in an unprecedented way.

On a broader perspective, our view of the world is undergoing a change. Irreversible evolution in natural processes is becoming a dominant theme. Since the dawn of modern science, our view of nature was dominated by the search for static immutable laws and the rise of the mechanistic picture. From Newton to Maxwell and Einstein, time was reduced to a parameter in the dynamical description of the world: irreversibility was only an illusion. This position is no longer defensible. Karl Popper expressed his view in these words: "The *reality of time and change* seems to me the crux of realism."[21] The current trends in science that developed in the last few decades seem to concur with this view.

Acknowledgments

A part of this work was sponsored by the Robert A. Welch Foundation of Houston, Texas.

I want to warmly thank Dr. D. Kondepudi for his kind help in the preparation of this article and for numerous discussions. I also want to thank Prof. P. Glansdorff for a discussion of the historical background of De Donder's work.

References

1. A. Eddington, *Mathematical Theory of Relativity*, Cambridge University Press, Cambridge, 1924.

2. V. A. Fock, *Space Time and Gravitation*, Macmillan, New York, 1964.

3. J. W. Gibbs, *Collected Works*, Longmans, Green, New York, 1928.

4. T. DeDonder, *L'Affinité*, Gauthier-Vallars, Paris, 1927.

5. R. Alpher and R. Herman, *Nature (London)*, **162**, 774 (1948); *Phys. Rev.*, **75**, 1089 (1949). See also *Proc. Am. Phil. Soc.*, **119**, 325 (1975).

6. G. Nicolis and I. Prigogine, *Self-Organization in Nonequilibrium Systems*, Wiley, New York, 1977.

7. P. Glansdorff and I. Prigogine, *Thermodynamics of Structure, Stability and Fluctuations*, Wiley, London, 1971.

8. J. C. Roux et al., in *Nonlinear Problems: Present and Future*, Proc. of Conf. at Los Alamos, A. R. Bishop, ed., North-Holland, Amsterdam, 1982; J. S. Turner et al., *Phys. Lett.*, **85A**, 9 (1981).

9. P. Ortoleva and J. Ross, *J. Dev. Biol.*, **34**, F19 (1973).

10. L. F. Jaffe and R. Nuccitelli, *Annu. Rev. Biophys. Bioeng.*, **6**, 445–476 (1977).

11. R. Larter and P. Ortoleva, *J. Theor. Biol.* **88**, 599 (1981).

12. O'Keefe, private communication.

13. P. Ortoleva et al., in *Instabilities, Bifurcation and Fluctuations in Chemical Systems*, L. E. Reichl and W. Schieve, eds., University of Texas Press, Aus. Texas, 1982; C. S. Hoase et al., *Science*, **209**, 272–274 (1980).

14. D. K. Kondepudi and I. Prigogine, *Physica*, **107A**, 1–24 (1981).

15. G. Nicolis and I. Prigogine, *Proc. Natl. Acad. Sci. USA*, **78**, 659–663 (1981); also see article by G. Nicolis in this volume.

16. G. Nicolis, in *Stochastic Nonlinear Systems in Physics, Chemistry and Biology*, L. Arnold and R. Lefever, eds., Springer-Verlag, Berlin and New York, 1981, pp. 44–52.

17. D. Walgraef et al., "Nonequilibrium Phase Transition and Chemical Instabilities" in *Adv. Chem. Phys.*, **49**, 311 (1982); also, A. Nitzan and P. Ortoleva, *Phys. Rev.*, **21A**, 1735–1755 (1980).

18. P. W. Anderson, in *Order and Fluctuations in Equilibrium and Nonequilibrium Statistical Mechanics* (XVIIth Solvay Conference on Physics), G. Nicolis, G. Dewel, and J. W. Turner, eds., Wiley, New York, 1981.

19. M. Malek Mansour et al., *Ann. Phys.*, **131**, 283–313 (1981).

20. D. H. Sattinger, *Arch. Rat. Mech. Anal.*, **66**, 31–42 (1977).

21. K. Popper, *Unended Quest*, Open Court Publishing Company, La Salle, Illinois, 1974, p. 129.

ATMOSPHERIC CHEMISTRY

MARCEL NICOLET

Géophysique Externe, Université Libre de Braxelles
and
Ionosphere Research Laboratory, The Pennsylvania State University

I. INTRODUCTION

The necessary starting point for any study of the chemistry of a planetary atmosphere is the dissociation of molecules, which results from the absorption of solar ultraviolet radiation. This atmospheric chemistry must take into account not only the general characteristics of the atmosphere (constitution), but also its particular chemical constituents (composition). The absorption of solar radiation can be attributed to carbon dioxide (CO_2) for Mars and Venus, to molecular oxygen (O_2) for the Earth, and to methane (CH_4) and ammonia (NH_3) for Jupiter and the outer planets.

So far as the terrestrial atmosphere is concerned, the dissociation of molecular oxygen by ultraviolet radiation leads to the formation of ozone (O_3) in dry air, which consists essentially of 78% nitrogen, 21% oxygen, and 1% argon. It follows that the characteristic element in the chemistry of the terrestrial atmosphere is oxygen in its three forms: O_2, O, and O_3, since nitrogen is not dissociated by the radiations absorbed by oxygen and ozone.

Terrestrial atmospheric chemistry varies in certain respects in the different regions of the homosphere (below 100 km), but it is characterized by the constant relationship between the three principal constituents of the atmosphere: N_2, O_2, and Ar. Within the homosphere, it is necessary to distinguish between the chemistries in three regions, in each of which there are differences in the relative abundance of trace gases consisting of, for example, molecules containing hydrogen, nitrogen, and the halogens. These regions are the troposphere (at ≤ 9 km over the poles, and ≤ 17 km over the equator), the stratosphere (about ≤ 50 km), and the mesosphere (about ≤ 85 km). The chemistry of these three regions can be studied separately if one takes account of possible interactions at the boundaries between the regions.

II. ABSORPTION OF SOLAR RADIATION

In order to study the photochemical action of solar radiation on tropospheric, stratospheric, and mesospheric constituents, the solar spectrum must be divided in various ranges.[1] The radiation at wavelengths less than 100 nm, which is absorbed by nitrogen and oxygen in the thermosphere above 100 km, leads essentially to ionization processes and is, therefore, not considered there. Only X-rays of wavelengths less than 1 nm can penetrate into the atmosphere below 100 km, and lead indirectly to the dissociation of molecular constituents. Nevertheless, their principal role is the photoionization in the D region of the ionosphere below 100 km where the solar line Lyman-α at 121.6 nm ionizes the nitric oxide molecule, NO.

The spectral range of wavelengths greater than 100 nm leads to the photodissociation of molecular oxygen. The principal absorption continuum (Schumann–Runge continuum) at wavelengths less than 175 nm can be neglected in the stratospheric and mesospheric chemistry of the homosphere since it is a thermospheric process occurring above 100 km. But, at shorter wavelengths, the solar line, Lyman-α at 121.6 nm, is situated in a so-called atmospheric window where the O_2 absorption cross section is relatively small. Such a radiation absorbed in the mesosphere varies strongly with solar activity and can, therefore, play an important role in the photodissociation processes of O_2, and also of H_2O and CO_2.

Another important spectral range, between 200 and 175 nm, is related to the O_2 Schumann–Runge band system which includes 18 bands, (2-0) to (19-0) subject to the predissociation processes, particularly in the mesosphere.

From 200 to 242 nm, the O_2 absorption, which is related to the Herzberg continuum with low absorption cross section, occurs in the stratosphere. In this spectral region, the ozone absorption must be introduced since the O_3 Hartley band is characterized by high values of the absorption cross section between 200 and 300 nm. The simultaneous absorption by O_2 and O_3 must be considered in the stratosphere, whereas the O_3 absorption is practically negligible in the mesosphere since the total number of absorbing ozone molecules is small.

At wavelengths greater than 310 nm, the Huggins bands correspond to the limit of the O_3 ultraviolet absorption, and in the visible region (410–850 nm) the Chappuis bands play an important role leading to the O_3 photodissociation in the lower part of the atmosphere, troposphere, and lower stratosphere.

III. THE PURE OXYGEN ATMOSPHERE

Before introducing the atmospheric problem in its various aspects, it is useful to begin the study with the photodissociation of molecular oxygen, and the subsequent reactions, in an oxygen atmosphere without any active chemical action on the part of other constituents (Fig. 1).

The radiation of wavelengths less than 242 nm is absorbed by molecular oxygen and leads to its photodissociation:

$$(J_2); \quad O_2 + h\nu \ (\lambda < 242 \text{ nm}) \rightarrow O + O \tag{1}$$

The peak of the O_2 photodissociation occurs in the stratosphere (near 35 km for an overhead sun) where the total number of O_2 molecules photodissociated is of the order of 10^7 cm^{-3} sec^{-1}. Below the ozone peak (≤ 25 km) the photodissociation rate decreases rapidly, particularly when the solar zenith angle increases. Below 20 km, the atomic oxygen production becomes very small and there is no atomic oxygen production in the troposphere by the O_2 photodissociation. The ozone photodissociation is the result of the absorption of solar radiation in the visible and the ultraviolet:

$$(J_3); \quad O_3 + h\nu \ (\lambda < 1 \ \mu\text{m}) \rightarrow O_2 + O \tag{2}$$

Because of absorption in the three spectral ranges covered by the Hartley, Huggins, and Chappuis bands, the rate of photodissociation of O_3 is a frequency of 10^{-2} sec^{-1} in the mesosphere, and greater than 10^{-4} sec^{-1} in the stratosphere and troposphere.

The recombination of atomic oxygen in the presence of a third body M (only N_2 and O_2)

$$(k_1); \quad O + O + M \rightarrow O_2 + M \tag{3}$$

cannot be neglected in the mesosphere above 70 km, but plays no practical role in the study of stratospheric processes. The reaction that does play a role at all altitudes is the association of oxygen atoms with oxygen molecules which leads to the formation of ozone:

$$(k_2); \quad O + O_2 + M \rightarrow O_3 + M \tag{4}$$

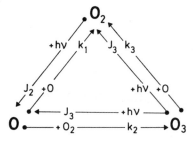

Fig. 1. Reaction scheme in an oxygen atmosphere with the photodissociation of O_2 and O_3.

Finally, ozone molecules and oxygen atoms react together and lead to the re-formation of oxygen molecules and the final destruction of ozone:

$$(k_3); \quad O + O_3 \rightarrow 2\,O_2 \tag{5}$$

Thus, the equations governing the rate of change of the concentrations of ozone, $n(O_3)$, and of atomic oxygen, $n(O)$, are

$$\frac{\partial n(O)}{\partial t} + \text{div}[n(O)\,w(O)] + 2k_1 n(M)\,n^2(O)$$

$$+ k_2 n(M)n(O_2)n(O) + k_3 n(O_3)n(O) = 2\,n(O_2)J_2 + n(O_3)J_3 \tag{6}$$

and

$$\frac{\partial n(O_3)}{\partial t} + \text{div}[n(O_3)\,w(O_3)] + n(O_3)J_3 + k_3 n(O)n(O_3)$$

$$= k_2 n(M)n(O_2)n(O) \tag{7}$$

where $\text{div}[nw]$ is the transport term that must be introduced when atmospheric exchanges (transport phenomena) are more significant than chemical processes (reactions). This happens above 80 km for atomic oxygen and below 35 km for ozone.

When photochemical equilibrium conditions can be applied, equations (6) and (7) are written in a practical form

$$n_*^2(O_3) = \frac{J_2}{J_3}\,n(M)n^2(O_2)\frac{k_2}{k_3} \tag{8}$$

and

$$n_*(O) = \frac{n(O_3)J_3}{k_2 n(M)n(O_2)} \tag{9}$$

These two equations are the conventional equations of ozone and atomic oxygen in an oxygen atmosphere where other constituents play no chemical role.

These basic photochemical and chemical processes for the formation and removal of atmospheric ozone were proposed by Chapman[2] 50 years ago.

The atmospheric conditions and the rates of the various reactions for which measured laboratory values are available indicate that the maximum rate of photodissociation of oxygen occurs in the stratosphere, and results in a maximum concentration of ozone also in the stratosphere at about 25 km.[3] Detailed studies show that the ozone is in photochemical equilibrium at levels above that of its maximum concentration. In other words, the production of ozone by solar ultraviolet radiation is automatically compensated by the re-formation of oxygen molecules in the upper stratosphere where equations (8) and (9) can be applied.

It has been known for about 50 years[4] that the annual variations in ozone do not correspond to these of the solar radiation depending on the latitude and the season. The behavior of ozone is characterized by a maximum in spring and a minimum in autumn; also there is more ozone at high than at low latitudes. This behavior shows that the chemical reactions in question are slow, in comparison with transport phenomena, in the lower stratosphere below 25 km.

Thus, when studying atmospheric chemistry, it is necessary always to take into account the vertical and horizontal movements in the atmosphere, as well as the conditions controlling those chemical reactions that do not spontaneously lead to photochemical equilibrium. These conditions are applicable not only to ozone in the lower stratosphere, but also to atomic oxygen in the upper mesosphere above 75 km. In fact, equation (4) shows that, with increasing height, the formation of O_3 becomes less and less important because of the decrease in the concentration of O_2 and N_2. Above 60 km the concentration of atomic oxygen exceeds that of ozone, but it is still in photochemical equilibrium up to 70 km. However, at the mesospause (85 km), it is subject to atmospheric movements, and its local concentration depends more on transport than on the rate of production.

A comparison of the expected and actual concentrations of ozone in the stratosphere shows that the theoretical values are greater than those observed. This difference can be attributed to reactions that would not occur in a pure oxygen atmosphere, but which are important in the real atmosphere because of the presence of various minor constituents.

IV. ACTION OF MINOR CONSTITUENTS

Experimental studies have shown that ozone can be removed as a result of cyclic catalytic reactions of the following type:

$$k_X; \quad X + O_3 \rightarrow XO + O_2$$

$$k_{XO}; \quad XO + O \rightarrow X + O_2$$

in which X denotes an atom or a molecule such as hydrogen, nitric oxide, the hydroxyl and perhydroxyl radicals, and even halogen atoms (Fig. 2).

Thus a number of catalytic reactions are associated with the re-formation of O_2 already indicated in equation (5). In other words, a halogen, or any other atom which attacks ozone, always reappears as a result of the reaction between its oxide and atomic oxygen, and hence there is a permanent cycle leading to the destruction of ozone. In atmospheric chemistry, therefore, it is important to find out how these constituents appear, and to assess their importance.

In fact, the photoequilibrium equation (8) becomes

$$n_*^2(O_3) = \frac{J_2}{J_3} n(M)n^2(O_2) \frac{k_2}{k_{3A}} \tag{8a}$$

where

$$k_{3A} = k_3 \left[1 + \frac{\sum k_{XO}n(XO)}{k_3 n(O_3)} \right] \tag{8b}$$

shows the specific action of the various oxides which may play role in the stratospheric ozone.

The existence, up to high levels of the atmosphere, of water vapor (H_2O), methane (CH_4), and molecular hydrogen (H_2) is well known (Fig. 3). In the mesosphere, H_2O and CH_4 are photodissociated in processes which result in the appearance of atomic hydrogen and the hydroxyl rad-

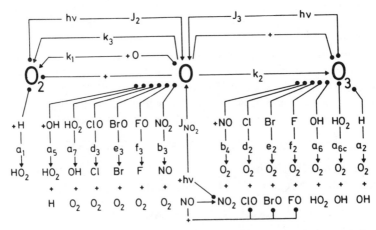

Fig. 2.   General reaction scheme in the atmosphere in which are simultaneously involved the chlorine, . . . , nitrogen, and hydrogen radicals (atoms or molecules) related to the production or loss of odd oxygen in the stratosphere.

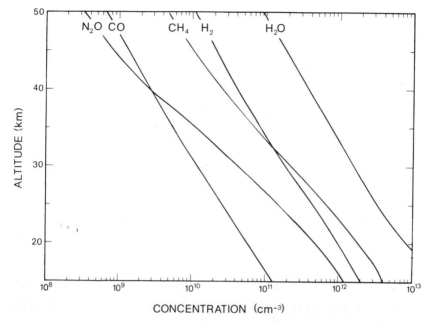

Fig. 3. Vertical distribution of the concentration of various minor constituents: water vapor, H_2O; methane, CH_4; molecular hydrogen, H_2; nitrous oxide, N_2O; and carbon monoxide, CO.

ical such as the following:

$$H_2O + h\nu \ (\lambda < 200 \text{ nm}) \rightarrow OH + H \qquad (10a)$$

$$\rightarrow O + H_2 \qquad (10b)$$

$$CH_4 + h\nu \ (\text{Lyman-}\alpha) \rightarrow CH_2 + H_2 \qquad (10c)$$

However, in the stratosphere, where H_2O is not photodissociated, it is necessary to consider a reaction involving the ozone photodissociation. This is a process in which an excited oxygen atom $O(^1D)$ reacts with a molecule to produce H and OH (Fig. 4) as follows:

$$O(^1D) + H_2O \rightarrow OH + OH \qquad (11a)$$

$$O(^1D) + CH_4 \rightarrow CH_3 + OH \qquad (11b)$$

$$O(^1D) + H_2 \rightarrow H + OH \qquad (11c)$$

The origin of the excited oxygen atom is ozone, but its concentration in the stratosphere is only 100–200 cm^{-3}, as compared with 10^{18} cm^{-3} for N_2 and O_2. In fact the absorption in the Hartley band, at wavelengths less than 310 nm, leads to the production of oxygen molecules and atoms, both in excited states:

$$O_3 + h\nu(\lambda < 310 \text{ nm}) \rightarrow O_2(^1\Delta_g) + O(^1D). \qquad (12)$$

Because of the additional energy in the excited atom $O(^1D)$, the rapid reactions (11a)–(11c) are possible; they would not be possible for the atom in the ground state, $O(^3P)$, at least at atmospheric temperatures.

It is necessary also to discover the identity of other constituents that might be involved in such dissociation reactions, and an important case is the transformation of nitrous oxide, N_2O, into nitric oxide, NO:

$$N_2O + O(^1D) \rightarrow NO + NO \qquad (13a)$$

or

$$\rightarrow N_2 + O_2 \qquad (13b)$$

Reaction (13a) is the principal production process of nitric oxide in the stratosphere.[5]

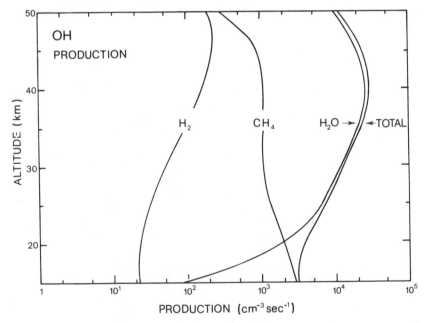

Fig. 4.   Vertical distribution of the OH radical production resulting from H_2O, CH_4, and H_2.

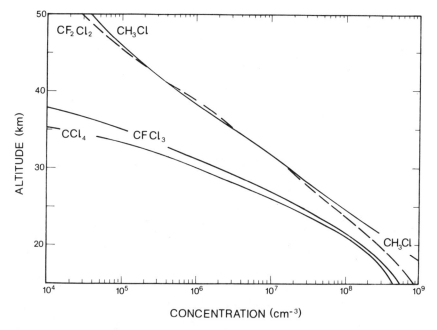

Fig. 5. Vertical distribution of the concentration of chloromethanes: CH_3Cl, methyl chloride; CF_2Cl_2, difluorodichloromethane; $CFCl_3$, fluorotrichloromethane; and CCl_4, carbon tetrachloride.

Finally, it is necessary to know the abundance in the atmosphere of minority constituents such as H_2O, N_2O, and so on (Figs. 3 and 5).

V. CONSTITUENTS CONTAINING HYDROGEN, AND THE CHEMISTRY OF THE MESOSPHERE

The importance for stratospheric chemistry of reactions taking place in an atmosphere containing oxygen and hydrogen has been known for about 30 years.[6] The presence of atomic hydrogen gives rise to the following reactions:

$$(a_1) \; H + O_2 \, (+ \, M) \rightarrow M + HO_2 \tag{14}$$

$$(a_2) \; H + O_3 \rightarrow O_2 + OH^* \tag{15}$$

$$(a_3) \; H + O_3 \rightarrow O + HO_2^* \tag{16}$$

Bates and Nicolet recognized the extreme importance of reaction 15 because it leads to the formation of an excited molecule OH^* which returns to its ground state with the emission of infrared radiation near 1 μm. This radiation is by far the strongest in the night airglow and is due to the reaction of ozone with atomic hydrogen produced by the photo-dissociation of water vapor. The presence of H and OH results in the catalytic cycle shown below

$$(a_2); \quad H + O_3 \rightarrow O_2 + OH \tag{15}$$

$$(a_5); \quad OH + O \rightarrow O_2 + H \tag{17}$$

Another cycle similar to (17) is also observed:

$$(a_7); \quad HO_2 + O \rightarrow O_2 + OH \tag{18}$$

This reaction with reactions (17) and (15), together constitute a continual attack on atmospheric ozone. Finally, in the mesosphere, the reaction of H with HO_2 must be taken into account:

$$H + HO_2 \rightarrow O + H_2O \tag{19a}$$

$$\rightarrow OH + OH \tag{19b}$$

$$\rightarrow O_2 + H_2 \tag{19c}$$

The last of these reactions is an important source of molecular hydrogen in the mesosphere.

The very light gases, atomic and molecular hydrogen, which have their origin in the water vapor in the mesosphere, are subject to vertical diffusion and are transported to the upper levels of the atmosphere, where they form the terrestrial hydrogen corona, and from where finally they escape continually into interplanetary space.

Although other factors are involved, it can be said that, from a practical point of view, mesospheric chemistry can be summed up as a series of interactions between the different forms of oxygen (O_2, O, and O_3), and of hydrogen and the hydroxyls (H_2, H, OH, and HO_2).

VI. OXYGEN AND NITROGEN COMPOUNDS AND STRATOSPHERIC CHEMISTRY

The importance of the oxides of nitrogen arises from the catalytic cycle involving NO and NO_2,[5,7,8]

$$(b_3); \quad NO_2 + O \rightarrow O_2 + NO \tag{20}$$

$$(b_4); \quad NO + O_3 \rightarrow O_2 + NO_2 \tag{21}$$

This cycle would lead to considerable destruction of ozone if it were not for the moderating influence of the very rapid photodissociation of NO_2 by solar radiation at wavelengths less than 400 nm:

$$NO_2 + h\nu \; (\lambda < 400 \text{ nm}) \rightarrow NO + O \tag{22}$$

In a sense, this reaction represents the production of ozone through the generation of atomic oxygen, which can associate with O_2 to form O_3.

On the other hand the concentration of NO and NO_2 in the stratosphere, originating continually from N_2O [equation (13a)], is limited by other reactions, in particular the one that leads to the formation of nitric acid:

$$OH + NO_2 \; (+ M) \rightarrow M + HNO_3 \tag{23}$$

Although the nitric acid molecule is subject to various reactions and to photodissociation, nevertheless it remains, and it becomes the most important of the molecules containing NO (HNO_4, N_2O_5, NO_3, . . .) in the lower stratosphere. However, it cannot accumulate because it crosses the tropopause into the troposphere, where it rapidly disappears because of its solubility in water. Thus, if N_2O is the source of the nitrogen oxides in the stratosphere, nitric acid is the sink that prevents their accumulation beyond certain limits. But it is now known that the sequence of reactions (20), (21), and (22) results in a lower concentration of stratospheric ozone than would be possible in a pure oxygen atmosphere.

Finally, it is necessary to take account of the interaction between NO and HO_2:

$$HO_2 + NO \rightarrow OH + NO_2 \tag{24}$$

This reaction not only restores the hydroxyl, but it also results in the formation of nitrogen dioxide which, because of its almost immediate photodissociation, leads to the liberation of an oxygen atom and to the formation of ozone. This last reaction plays an important role in the lower stratosphere where the photodissociation of molecular oxygen is no longer important.

VII. THE HALOGEN COMPOUNDS AND STRATOSPHERIC CHEMISTRY

The halogen family, fluorine, chlorine, bromine, and iodine has only recently been introduced into atmospheric chemistry.[9] The catalytic cycle, with an atom such as Cl in the presence of its monoxide ClO, can be applied giving, in effect:

$$(d_2); \quad Cl + O_3 \rightarrow O_2 + ClO \tag{25}$$

and

$$(d_3); \quad ClO + O \rightarrow O_2 + Cl \tag{26}$$

However, it is appropriate to introduce other possible reactions, such as that by which NO restores Cl and thus limits the absolute importance of reactions (25) and (26). This reaction has, at the same time, the effect of creating NO_2, which, as indicated earlier, leads to the restoration of atomic oxygen.

Among other reactions, it is worth mentioning the formation of hydrogen chloride in a reaction with methane, molecular hydrogen, or other constituents such as formaldehyde, H_2CO, or hydrogen peroxide, H_2O_2. An example is

$$Cl + CH_4 \rightarrow CH_3 + HCl \tag{27}$$

which is followed by

$$HCl + OH \rightarrow H_2O + Cl \tag{28}$$

Because of the reaction of HCl with OH, its production in the stratosphere is restricted. It must be remembered that HCl is the stratospheric sink for chlorine compounds because, after crossing the tropopause and entering the troposphere, it dissolves in water and disappears.

The atmospheric chemistry of bromine can be regarded as similar to that of chlorine. As far as HF is concerned, it does not react with hydroxyl; since it persists, it limits the concentration of the atom F and its oxide FO. Hence, HF is the sink for fluorine in the stratosphere, before it disappears in the troposphere.

VIII. SOURCES AND SINKS OF HYDROGEN, NITROGEN, AND HALOGEN COMPOUNDS

Although the troposphere has the characteristic of containing a high relative concentration of water vapor (10^{-5}–10^{-2}), the stratosphere is dry and the water vapor concentration is only a few parts in a million. However, the oxidation of methane by hydroxyl radical must be introduced:

$$CH_4 + OH \rightarrow CH_3 + H_2O \qquad (29)$$

This leads finally to H_2 and H_2O, and to CO which als'o reacts with OH to give carbon dioxide:

$$CO + OH \rightarrow H + CO_2 \qquad (30)$$

In the troposphere, OH is produced mainly by reaction (11a) and it is responsible for the oxidation of many constituents such as methyl chloride, chloroform, and a certain number of chlorofluoromethanes.

Methane is a naturally occurring gas that originates during the decomposition of organic matter under certain conditions. It appears in damp areas such as marshes and rice fields, and during the intestinal fermentation particularly of ruminant animals. Various studies lead to the methane budget shown in Table I.[10] Although the atmosphere contains about 4 gigatonnes of methane, it is estimated that this mass could be accumulated in only 5 or 10 years. The destruction of methane must be due to the reaction (29), and it is clearly possible that man may be able to influence the concentration of methane in the atmosphere through the use of processes involving microbes.

The study of the oxides of nitrogen in the stratosphere reduces to a consideration of the origin of nitrous oxide. The source of N_2O has been identified; after the fixation of nitrogen by biological or artificial processes, or in the atmosphere, it is transformed, in the soil, into nitrates; these are partly broken down by bacterial action into the gaseous products N_2 and N_2O which escape into the atmosphere (Table II). The lifetime of N_2O is sufficiently long in the troposphere to allow it to reach the stratosphere; there, before being dissociated into N_2 and O, it reacts with oxygen [process (13a)] to produce NO.

If the proportion of N_2O to N_2 in the denitrification process is of the order of 1/20, it can be seen that the quantity of N_2O being produced is

TABLE I. Global CH_4 Budget
(CH_4 Mt yr^{-1})

Source	Min	Max
Natural wetlands	175	300
Rice paddy fields	150	300
Enteric fermentation	75	200
	400	800

TABLE II. Global Nitrous Oxide Budget

Source of N_2 fixation	Mt yr^{-1}		
Natural processes			
Agricultural land	90		
Forested and unused land	60		
Industrial processes			
Fertilizers	40		
Combustion	20		
Lighting	10		
Ocean	(1)		
Total	220	175	150
	(in 1975)	(in 1950)	(in 1850)
N_2O formation (5%)	11	9	7,5

of the order of 10–15 megatonnes per year. Since the total atmospheric content of N_2O is about 1.5 gigatonnes, about 100 years would be required to produce such a quantity. It may be concluded that the industrial fixation of nitrogen could have a long-term effect, for the amounts involved have continued to increase since 1950, and it is estimated that 100–200 megatonnes per year may be produced in about the year 2000. The industrial fixation of nitrogen, which already accounts for 20% of the total, therefore, could disturb the equilibrium of the stratosphere if it were to increase beyond certain limits (for details and references, see ref. 11).

The halocarbons, which are not destroyed in the troposphere by reactions with hydroxyl, pass into the stratosphere where they are photodissociated to liberate chlorine atoms which attack ozone. Only one of them is of natural origin, methyl chloride CH_3Cl, but there are also several industrial products, especially carbon tetrachloride, CCl_4, trichlorofluoromethane, $CFCl_3$, and dichlorodifluoromethane. Methyl chloride (Table III) has a natural marine origin (for details, see ref. 12), but it is certainly present also in the smoke produced when polyvinyl and other products containing chlorine are burnt. In addition, it is produced naturally not only in forest fires, but also in tropical agriculture based on the cultivation

TABLE III. Sources of Methyl Chloride

Source	Mt yr^{-1}
Microbes	?
Urban anthropogenic	0.15–0.6
Combustion of vegetation	0.4 –1
Marine	1 –8
Total	1.6 –9.6

TABLE IV. Production and Dispersion
(Mt yr^{-1}) of Halocarbons in 1973

	CF_2Cl_2	$CFCl_3$	CCl_4
Production	0.42	0.34	1.0
Dispersion	0.36	0.28	0.1

of burnt-out forest lands, and during the smouldering of carbonaceous material. Table IV brings together the rates of industrial production and atmospheric dispersion of CF_2Cl_2, $CFCl_3$, and CCl_4.

Carbon tetrachloride, CCl_4, is generally used in the production of other substances, whereas the halocarbons 13 and 12 are in daily use in sprays and refrigerators. The annual rate of production of $CFCl_3$ was 1000 tonnes in 1947; it rose about to 35,000 tonnes in 1957, 150,000 tonnes in 1967, and to at least 300,000 tonnes in 1979 (for details see ref. 13). If the dispersion of these halocarbons in the atmosphere continues to increase, then the annual rate at which they are produced is so great that it cannot fail to have an effect on the accumulated chlorine in the stratosphere, and on the stratospheric ozone.

References

1. M. Nicolet, Solar UV radiation and its absorption in the mesosphere and stratosphere, *Pure Appl. Geophys.*, **118**, 3 (1980).

2. S. Chapman, A theory of upper atmospheric ozone, *Mem. Roy. Meteorol. Soc.*, **3**, 103 (1930).

3. H. U. Dütsch, Atmospheric ozone and ultraviolet radiation, in *Climate of the Free Atmosphere of World Survey of Climatology*, Vol. 4, H. Landsberg, ed., pp. 383–432, Elsevier, Amsterdam, 1969.

4. M. Nicolet, Etude des réactions chimiques de l'ozone dans la stratosphère, Inst. Roy. Météorol. Belgium, Brussels, 1980, 536 pp.

5. M. Nicolet, Aeronomic reactions of hydrogen and ozone, Aeronomica Acta A no. 79, 1970, and in *Mesospheric Models and Related Experiments*, Fiocco, ed., pp. 1–57, D. Reidel, Dordrecht, the Netherlands, 1971.

6. D. R. Bates and M. Nicolet, The photochemistry of atmospheric water vapor, *J. Geophys. Res.*, **55**, 301 (1950).

7. P. J. Crutzen, The influence of nitrogen oxides on the atmospheric ozone content, *Q. J. Roy. Meteorol. Soc.*, **96**, 320 (1970).

8. H. S. Johnston, Reduction of stratospheric ozone by nitrogen oxide catalysts from supersonic transport exhaust, *Science* **173**, 517 (1971).

9. F. S. Rowland and M. J. Molina, Chlorofluoromethanes in the environment, *Rev. Geophys. Space Phys.*, **13**, 1 (1975).

10. T. M. Donahue, The atmospheric CH_4 budget, *Proc. NATO Advanced Study Institute on Atmospheric Ozone: Its Variation and Human Influences*, Aldeia das Açoteias,

Portugal, M. Nicolet, dir., and A. A. Aikin, ed., U.S. Dept. Trans., Report No. FAA-EE-80-20, p. 301, 1980.

11.   M. B. McElroy, Sources and sinks of nitrous oxide, *Proc. NATO Advanced Study Institute on Atmospheric Ozone: Its Variation and Human Influences*, Aldeia das Aço-teias, Portugal, M. Nicolet, dir., and A. A. Aikin, ed., U.S. Dept. Trans., Report No. FAA-EE-80-20, p. 345, 1980.

12.   A. J. Watson, J. E. Lovelock, and D. H. Stedman, The problem of atmospheric methyl chloride, *Proc. NATO Advanced Study Institute on Atmospheric Ozone: Its Variation and Human Influences*, Aldeia das Açoteias, Portugal, M. Nicolet, dir., and A. A. Aikin, ed., U.S. Dept. Trans., Report No. FAA-EE-80-20, p. 345, 1980.

13.   J. P. Jesson, Release of industrial halocarbons and tropospheric budget, *Proc. NATO Advanced Study Institute on Atmospheric Ozone: Its Variation and Human Influences*, Aldeia das Açoteias, Portugal, M. Nicolet, dir., and A. A. Aikin, ed., U.S. Dept. Trans., Report No. FAA-EE-80-20, p. 373, 1980.

COMMENTARY: OBSERVATIONAL ASPECTS RELATED TO THE CHEMICAL EVOLUTION OF OUR ATMOSPHERE

R. ZANDER

Institut d' Astrophysique, Université de Liège,
Liège-Ougrée, Belgium

Following the report on "The Chemistry of the Atmosphere," it appears urgent to intensify sustained observational work in order to establish facts about any eventual evolution of our global environment. Such facts need to be gathered not only for physical parameters (temperature, albedo, variations of the solar ultraviolet irradiance below 3200 Å, diffusion coefficients, aerosols, etc.) but also for a growing number of chemical species whose telluric concentrations are ultimately controlling the state of that environment. In 1960, a dozen molecules were known to exist in our atmosphere; by 1980, 25 more species have been added to these and experimenters are asked now to look for another 40 molecules likely to play a role in the complex aeronomical scheme outlined by Professor M. Nicolet.

Our purpose here is not to go into the details of the subject raised but to stress major aspects which have emerged recently and which need to be covered by future coordinated observational programs.

From now on, it will for instance be necessary to discern among compositional variability, trends, and pollution.

Variability may be defined as reflecting fluctuations in the atmosphere, of natural origin, with both temporal and spatial scales; examples are diurnal, seasonal, solar activity-related variations; impulsive events such as volcano eruptions and solar proton events; fluctuations linked to some peculiar meteorological conditions, for example, intense cyclonic activities and jet streams. Variability by itself is a whole program to be conducted ideally on a four-dimensional basis (latitude, longitude, altitude, and time) by space vehicles, for example, satellites or from the space shuttle. This area of research is certainly the most urgent one to be developed during the next decade, as it relates to fundamental interactions among chemistry, radiation, and dynamics in the atmosphere, mainly in the 15–75 km altitude range.

Trends are characterized by long and smooth periodic changes that apply to the global atmosphere (occasionally, lag times may exist between either hemispheres). Until 30 years ago, trends would only have been evoked within the context of a slow, natural evolution of the earth's environment and of its average climate. However, since about 1950, we know of man's impact on the CO_2 burden in the atmosphere, leading to an actual increase of about 0.4% per year.[1] Furthermore, based on the consequences theoretically predicted during the last decade about the erosion of the stratospheric ozone layer by the nitrogen (NO_x) and chlorine (ClO_x) catalytic cycles,[2,3] our atmosphere may be experiencing an accelerated evolution, which we need to assess experimentally; this applies by priority to the total ozone content of the atmosphere and it is going to require much better coordinated and more accurate O_3 measurements than over the last 20 years; anticipated destruction of ozone by either nuclear explosions during the 1950s[4] or by anthropogenic chlorinated species[5] could not be corroborated experimentally because of strong natural O_3 variability and too large observational errors, seldom better than $\pm$ 10%. Trends studies demand continuity in the observations and exceptionally good stability and accuracy of the techniques adopted. For short-lived species (e.g., OH, O_3 above 30 km), global coverage with altitude resolution will again be ideal; for long-lived ones, column density measurements to be carried out from a few selected ground stations may be satisfactory: the CO_2 monitoring program is a good example. We believe that other minor constituents such as N_2O, CH_4, CO, and H_2O (which were all discovered through solar observations from the ground) could as well be efficiently monitored through regular column density measurements from high-altitude mountains (H_2O may require better global coverage because of its high variability both in the troposphere and in the stratosphere).

Recently recognized *pollution* has brought up a series of new molecules whose accumulation in the atmosphere needs to be closely followed. Related observations ought to include the pollutants themselves (e.g., secular minor constituents such as NO, NO_2, N_2O, and CO whose burden is affected by man's activity, and new species such as $CFCl_3$, CF_2Cl_2, CCl_4, CH_3Cl, COS) as well as by-products which they lead to after photodissociation or/and chemical reactions in the upper stratosphere (e.g., HNO_3, HCl, HF, ClO, $ClONO_2$, HOCl); some of these are important sinking compounds which dampen the catalytic cycles.[6] In the pollution context, it is also primordial to establish the relative importance of all potential sources of pollution, natural and anthropogenic, most of which are on the ground.

Methods and techniques necessary for investigating the various fields of research mentioned here before exist: they are either remote or *in situ*, active or passive, in real time or delayed, from the ground, on-board airplanes, balloons, rockets, or satellites (soon from the space shuttle). Their specific advantages and prevailing fields of applications have often been demonstrated or confirmed at the occasion of observational programs carried out during the 1970s; a good review of these techniques may be found in ref. 7.

Most important in the near future will be

The coordination of observational programs, in order to obtain simultaneously data regarding interrelated constituents, for example, the NO_x or ClO_x group of species in the stratosphere.

The improvement of four-dimensional observations for global variability studies between 15 and 75 km altitude, with a special emphasis on the accuracy and sensitivity of the embarked experiments.

The study of the low atmosphere (including land and ocean exchanges with or through the atmosphere) which is the supply of most material found in the stratosphere.

The temporal continuity of measurements devoted to trends studies of the most stable and strongly absorbing molecules.

Although one may expect a substantial contribution from experiments on-board of future satellites and from the space shuttle working in remote modes, either by absorption measurements during solar occultations or by limb scanning emission observations (this last technique allows diurnal variations studies), it is clear that ground, airplane, and balloon observations are going to remain vital during the next decade. For instance, they will provide valuable support to space programs for data validation; they will also allow to test new instruments under relaxed timing before having these put on-board of expensive space platforms. Furthermore, the search of new species in our atmosphere needs to be carried out with the most powerful instruments existing and this is not yet the virtue of the space equipments.

We like to conclude these few observational aspects by mentioning that our group, under Professor M. Migeotte's heading, has contributed significantly to the investigation of the chemical composition of the earth's atmosphere through solar observational programs carried out since 1950 from the International Scientific Station of the Jungfraujoch located at 3580-m altitude in the Swiss Alps[8,9,10] and, since 1970, by infrared balloon observations.[11,12]

During the last years, ground monitorings of the column densities of
HF, HCl, CH_3Cl, and CH_4 were made from both the Jungfraujoch and
from the Kitt Peak Observatory, Tucson, Arizona, each time that observations were undertaken there by L. Delbouille and G. Roland. Figure 1
shows the results accumulated for HF whose column density can easily
be measured from the ground by high spectral resolution observations
near 2.5 μm[13]; the high mixing ratios observed occasionally are due to
strong exchanges between the stratosphere and the troposphere. The set
of data shown here suggests an increase of HF during the last years.
Another example is given in Fig. 2 which relates to CH_4 measurements
made in 1951[8] and in 1977. The equivalent widths per air mass were measured over the P(8) branch of the ν_3 band of CH_4 at 3.3 μm, on spectra
recorded under almost identical slant path conditions, but 26 years apart;
they all fit very well on a single curve, which is indicative of the stability
of CH_4 over that time span. In 1977, Malbrouck[14] had reached the same
conclusion when comparing 1975 results with earlier deductions by Nielsen and Migeotte.[15]

We have mentioned these two examples for raising the importance of

Making high spectral resolution infrared solar observations from the
ground (0.01–0.03 cm^{-1}), even about constituents which are predominant in the stratosphere.

Saving any old observational data for subsequent comparisons.

Repeating specific observations under conditions as identical as pos-

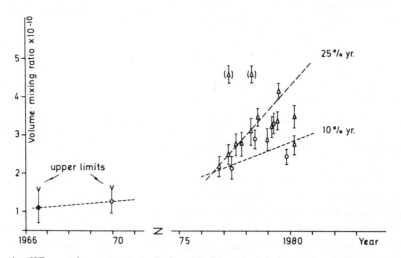

Fig. 1. HF ground measurements. Reduced to average mixing ratio above 20 km. $\triangle$, Jungfraujoch; $\bigcirc$, Kitt Peak.

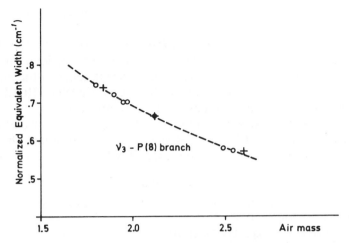

Fig. 2. Methane in the atmosphere at Jungfraujoch (3580-m altitude). +, 1951 (Migeotte, Neven, Swensson); O, 1977 (Delbouille, Roland).

sible, from the same site, which simplifies reduction procedures and intercomparisons.

From balloon observations carried out between 1974 and 1979, it has for instance been possible to show that the HF mixing ratio increases versus altitude, out to 38 km. Figure 3 displays recent results deduced about the HCl average mixing ratio (flights ULG-11 and ULG-15) and its

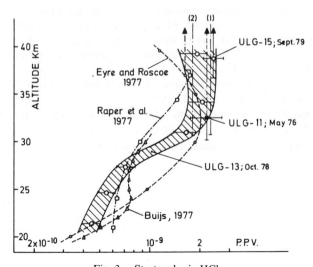

Fig. 3. Stratospheric HCl.

distribution obtained in 1978 (ULG-13) by subhorizontal sun observations near 3.3 μm with a resolution of 0.04 cm^{-1}.

Many other constituents, such as H_2O, CO_2, CH_4, N_2O, CO, HOCl, observed in 1979, are under investigation. On our next balloon campaign, we intend to measure simultaneously the concentrations of the most important chlorine species, for example, HCl, ClO, $ClONO_2$, HOCl, CCl_4, and CH_3Cl, through observations in the 3–12 μm region.

Our results reported here were obtained under contracts with the Belgian government, the Chemical Manufacturers Association, the Commission of the European Communities, and NASA.

References

1. C. A. Ekdahl and C. D. Keeling, in *Carbon and the Biosphere*, Atomic Energy Commission, Washington, D. C., 1973.

2. H. S. Johnston, *Science,* **173**, 517 (1971).

3. M. J. Molina and F. S. Rowland, *Nature (London)*, **249**, 810 (1974).

4. J. K. Angelli and J. Korshover, *Mon. Weather Rev.,* **101**, 426 (1973).

5. National Academy of Sciences, *Halocarbons: Effects on Stratospheric Ozone*, NAS, Washington, D.C., 1976.

6. N. D. Sze, M. B. McElroy, S. C. Wofsy, D. Kong, and R. Daesen, *Theoretical Models of Stratospheric Chemistry, Perturbations and Trace Gas Measurements*, Final report to Manufacturing Chemists Association, Washington, D.C., 1978.

7. Nat. Center for Atmosph. Res., *Instruments and Techniques for Stratospheric Research*, Atmosph. Technology series, No. 9, 1978.

8. M. Migeotte, L. Neven, and J. Swensson, *The Solar Spectrum from 2.8 to 23.7 Microns*, Mém. Soc. Royale Sci. Liège, Special Vol. No. 1, 1956.

9. L. Delbouille and G. Roland, *Photometric Atlas of the Solar Spectrum from λ7498 to λ12016 Å*, Mém. Soc. Royale Sci. Liège, Special Volume No. 4, 1963.

10. L. Delbouille, G. Roland, and L. Neven, *Photometric Atlas of the Solar Spectrum from λ3000 to λ10.000 Å*, Special Volume of the Institut d'Astrophysique de l'Univ. de Liège, Belgium, 1973.

11. R. Zander, Acad. Royale de Belgique, Cl. Sci., 5e serie, tome LVI, 729, 1970.

12. R. Zander, *Infrared Phys.,* **16**, 125 (1976).

13. R. Zander, G. Roland, and L. Delbouille, *Geophys. Res. Lett.,* **4**, 117, 1977.

14. R. Malbrouck, Acad. Royale de Belgique, Cl. Sci., 5e serie, tome LXIII, 773, 1977.

15. H. A. Nielsen and M. Migeotte, *Ann. Astrophys.,* **15**, 134, 1952.

THE PREBIOTIC SYNTHESIS
OF ORGANIC MOLECULES
AND POLYMERS

STANLEY L. MILLER

*Department of Chemistry, University of California at San Diego,
La Jolla, California*

I. INTRODUCTION

It is now generally accepted that life arose on the earth early in its history. The sequence of events started with the synthesis of simple organic compounds by various processes. These simple organic compounds reacted to form polymers, which in turn reacted to form structures of greater and greater complexity until one was formed which could be called living. This is a relatively new idea, first expressed clearly by Oparin[1] with contributions by Haldane,[2] Urey,[3] and Bernal.[4] It is sometimes referred to as the Oparin–Haldane or the Heterotrophic hypothesis.[5] Older ideas that are no longer regarded seriously include the seeding of the earth from another planet (panspermia), the origin of life at the present time from decaying organic material (spontaneous generation), and the origin of an organism early in the earth's history by an extremely improbable event. The latter process assumed that the earth was essentially the same as at present, except possibly for the absence of molecular oxygen, and so such an organism would have to be autotrophic. That is, it would have to synthesize all its organic compounds from CO_2, H_2O, and light.

It was Oparin's proposal that the first organisms were heterotrophic, that is, they utilized the organic compounds available in the environment. They still had to build proteins, nucleic acids, etc., but did not have to synthesize the amino acids, purines, pyrimidines, and sugars. Oparin as well as Urey also proposed that the earth had an atmosphere of CH_4, NH_3, H_2O, and H_2 and that organic compounds might be synthesized in such an atmosphere. On the basis of Urey's and Oparin's ideas, it was shown that amino acids could be synthesized in surprisingly high yield from the action of spark on a strongly reducing atmosphere.[6] There is now a large literature dealing with the prebiotic synthesis of organic com-

85

pounds and polymers, which is too extensive to review here, so only the highlights will be covered (for reviews, see refs. 7–9). In addition, the heterotrophic hypothesis and the detailed pathways of prebiotic synthesis place constraints on geological conditions on the primitive earth. Some of these constraints will be considered first.

A. The Time Available for the Origin of Life

The time period in which prebiotic synthesis of organic compounds took place is frequently misunderstood. The earth is 4.5×10^9 years old, and the earliest fossil organisms known, the Warrawoona microfossils and stromatolites, are 3.5×10^9 years old.[10] The difference is 1.0×10^9 years, but the time available for life to arise was probably shorter. It probably took a few hundred million years for organisms to evolve to the level of those found in the Warrawoona formation. In addition, if the earth completely melted during its formation, then the time available would be further shortened by the time needed for the earth to cool down sufficiently for organic compounds to be stable.

A period of say 0.5×10^9 years does not, in my opinion, present any problems. Many writers have stressed that many improbable events were required for the origin of life, and therefore much time was needed. I believe that too much emphasis has been placed on the need for time. Periods of 10^9 years are so far removed from our experience that we have no feeling or judgment as to what is likely or unlikely in them. If the origin of life took only 10^6 years I would not be surprised. It cannot be proved that 10^4 years is too short a period.

B. Melting of the Primitive Earth

There are many theories that seek to describe the events that took place during the condensation of cosmic dust and larger objects to form the earth. At one extreme, it is believed that the accumulation of material took place in less than 10^5 years; in this scheme gravitational energy was also released sufficiently rapidly to melt the entire earth, including material at its surface. At the other extreme, the accretion of material is believed to have been sufficiently slow, so that the gravitational energy released during accumulation was dissipated by radiation at a rate comparable to the rate of energy production. In this model the interior of the earth would have melted due to the effects of adiabatic compression and the decay of radioactive elements. In both models a molten core would have formed.

Whether the entire earth was ever molten or not does not greatly affect our discussion of the prebiotic synthesis of organic compounds. If the entire earth did melt, then all organic compounds, both in the interior and on the surface, would have been pyrolyzed completely to an equilibrium mixture consisting largely of CO_2, CO, CH_4, H_2, N_2, NH_3, and H_2O. All organic compounds synthesized in the solar nebula and reaching the surface of the earth intact would have been destroyed on a molten earth. When the earth had cooled down sufficiently, a crust would have formed. When the average temperature on the surface and in the atmosphere became low enough, organic compound would have been synthesized and, most important, would have accumulated. It is possible that some of the organic compounds synthesized in the solar nebula were brought to the earth with dust particles, meteorites, and comets and that they survived their impact with the atmosphere and with the earth's surface. It is not clear how much of a contribution these compounds made to the inventory of prebiotic organic compounds. Relatively unstable compounds, such as sugars and certain amino acids, cannot be accounted for in this manner since they would have decomposed before life could arise. There would have been a continuous addition of carbonaceous chondrites after the initial major accretion, but they would not result in the accumulation of unstable organic compounds. In any case, unstable organic compounds such as sugars have not been found in carbonaceous chondrites. Thus, for unstable compounds a continuous synthesis is required, which means synthesis in the atmosphere and oceans of the primitive earth.

C. Temperature of the Primitive Earth

The necessity to accumulate organic compounds for the synthesis of the first organism requires a low temperature earth, for otherwise the organic compounds would decompose. The half-lives for decomposition vary from several billion years for alanine at 25°C, a few million years for serine, 10^3 to 10^5 years for hydrolysis of peptides and polynucleotides, to at most a few hundred years for sugars. Since the temperature coefficients to these decompositions are large, the half-lives would be much less at temperatures of 50° or 100°C. Conversely, the half-lives would be longer at 0°C. The rates of hydrolysis of peptide and polynucleotide polymers, and of decomposition of sugars, are so large that it seems impossible that such compounds could have been accumulated in aqueous solution and have been used in the first organism, unless the temperature was low.

The temperature of the present ocean averages 4°C, with the surface waters somewhat warmer. Whereas the freezing point of water is 0°C, the freezing point of seawater is −1.8°C. Seawater solidifies almost completely at −21°C. The temperature of the primitive ocean is not known, but it can be said that the instability of various organic compounds and polymers makes a compelling argument that life could not have arisen in the ocean unless the temperature was below 25°C. A temperature of 0°C would have helped greatly, and −21°C would have been even better. At such low temperatures, most of the water on the primitive earth would have been in the form of ice, with liquid seawater confined to the equatorial oceans.

There is another reason for believing that life evolved at low temperatures, whether in the oceans or in lakes. All of the template-directed reactions that must have led to the emergence of biological organization take place only below the melting temperature of the appropriate organized polynucleotide structure. These temperatures range from 0°C, or lower, to perhaps 35°C, in the case of polynucleotide–mononucleotide helices.

The environment in which life arose is frequently referred to as a warm, dilute soup of organic compounds. I believe that a cold concentrated soup would have provided a better environment for the origins of life. At low temperatures the decomposition of organic compounds and polymers is slowed down greatly. Although at first sight low temperatures might seem to have been a disadvantage (chemical syntheses would have proceeded more slowly), in fact they may have been advantageous. It is the ratios of the rates of synthesis to the rates of decomposition which are important, rather than the absolute rates, if ample time is available. Since the temperature coefficients of the synthetic reactions are generally less than those for the decomposition reactions, low temperatures would have favored the synthesis of more complex organic compounds and polymers.

D. Oxygen in the Primitive Atmosphere

Molecular oxygen is usually assumed to have been absent from the primitive atmosphere. The first reason is that the major source, in the absence of O_2 evolving photosynthetic organisms, would have been the photodissociation of water in the upper atmosphere. This is not a large source of O_2, and it would react with Fe^{2+} and other reduced inorganic compounds. The second reason is that O_2 reacts relatively rapidly with organic compounds, especially in the presence of ultraviolet light. These and related arguments are so compelling that it does not seem possible that organic compounds remained in the primitive ocean for any length of time after large amounts of O_2 entered the earth's atmosphere. Organic compounds are present on the surface of the earth only because they are continuously being resynthesized by living organisms. Organic compounds occur below the surface of the earth, for example, in coal and oil, because there the environment is anaerobic. It appears certain then that substantial amounts of O_2 were absent from the earth's atmosphere during the period when organic compounds were synthesized and probably up to the time when the first organism evolved.

If the origin of life does not occur by the time a planet becomes oxidizing and the organic compounds are decomposed both thermally or by oxidation with O_2, then life can never arise on that plant, unless reducing conditions can be reestablished by some process.

E.   The Composition of the Primitive Atmosphere

There is no agreement on the constituents of the primitive atmosphere. It is to be noted that there is no geological evidence concerning the conditions on the earth from 4.5×10^9 years to 3.8×10^9 years since no rocks older than 3.8×10^9 years are known. Even the 3.8×10^9 year old Isua Rocks in Greenland are not sufficiently well preserved to infer details of the atmosphere at that time. Proposed atmospheres range from strongly reducing (CH_4, NH_3, H_2O, H_2) to nonreducing (CO_2, N_2, H_2O). The various proposed atmospheres and the reasons given to favor them will not be discussed here. As shown in the Section II the more reducing atmospheres favor the synthesis of organic compounds both in terms of yields and the variety of compound obtained. Some of the organic chemistry makes explicit predictions about atmospheric constituents. Such considerations cannot prove that the earth had a certain primitive atmosphere, but the prebiotic synthesis constraints should be a major consideration.

II.   PREBIOTIC SYNTHESES

A.   Energy Sources

A wide variety of energy sources has been utilized with various gas mixtures since the first experiments using electric discharges. The importance of a given energy source is determined by the product of the energy available and its efficiency for organic compound synthesis. Even though both factors cannot be evaluated with precision, a qualitative assessment of the energy sources can be made. It should be emphasized that a single source of energy or a single process is unlikely to account for all the organic compounds on the primitive earth. An estimate of the sources of energy on the earth at the present time is given in Table I.

The energy from the decay of radioactive elements was probably not an important energy source for the synthesis of organic compounds on the primitive earth since most of the ionization would have taken place in silicate rocks rather than in the reducing atmosphere. The shock wave energy from the impact of meteorites on the earth's atmosphere and sur-

TABLE I. Present Sources of Energy Averaged over the Earth[11]

Source	Energy $(cal\ cm^{-2}\ yr^{-1})$
Total radiation from sun	260,000
Ultraviolet light	
<3000 Å	3,400
<2500 Å	563
<2000 Å	41
<1500 Å	1.7
Electric discharges	4
Cosmic rays	0.0015
Radioactivity (to 1.0-km depth)	0.8
Volcanoes	0.13
Shock waves	1.1
Solar wind	0.2

face as well as the larger amount of shock waves generated in lightning bolts have been proposed as energy sources for primitive earth organic synthesis. Very high yields of amino acids have been reported in some experiments,[12] but it is doubtful whether such yields would be obtained in natural shock waves. Cosmic rays are a minor source of energy on the earth at present, and it seems unlikely that any increase in the past could have been so great as to make them a major source of energy.

The energy in the lava emitted at the present time is a significant but not a major source of energy. It is generally supposed that there was a much greater amount of volcanic activity on the primitive earth, but there is no evidence to support this. Even if the volcanic activity was a factor of 10 greater than at present, it would not have been the dominant energy source. Nevertheless, molten lava may have been important in the pyrolytic synthesis of some organic compounds.

Ultraviolet light was probably the largest source of energy on the primitive earth. The wavelengths absorbed by the atmospheric constituents are all below 2000 Å except for ammonia (< 2300 Å) and H_2S (< 2600 Å). Whether it was the most effective source of organic compounds is not clear. Most of the photochemical reactions would occur in the upper atmosphere, and the products formed would, for the most part, absorb the longer wavelengths, and so be decomposed before they reached the protection of the oceans. The yield of amino acids from the photolysis of CH_4, NH_3, and H_2O at wavelengths of 1470 and 1294 Å is quite low,[13] probably due to the low yields of hydrogen cyanide. The synthesis of amino acids by the photolysis of CH_4, C_2H_6, NH_3, H_2O, and H_2S mixtures by ultraviolet light of wavelengths greater than 2000 Å[14] is also a low yield synthesis, but the amount of energy is much greater in this region of the sun's spectrum. Only H_2S absorbs the ultraviolet light, but the photodissociation of H_2S results in a hydrogen atom having a high kinetic energy, which activates or dissociates the methane, ammonia, and water. This appears to be very attractive prebiotic synthesis. However, it is not clear whether a sufficient partial pressure of H_2S could be maintained in the atmosphere since H_2S photolyzed rapidly to elemental sulfur and hydrogen.

The most widely used sources of energy for laboratory syntheses of prebiotic compounds are electric discharges. These include sparks, semicorona, arc, and silent discharges with the spark being the most frequently used type. The ease of handling and high efficiency of electric discharges are factors favoring its use, but the most important reason is that electric discharges are very efficient in synthesizing hydrogen cyanide, whereas ultraviolet light is not. Hydrogen cyanide is a central intermediate in prebiotic synthesis, being needed for amino acid synthesis from the Strecker reaction, or by self-polymerization to amino acids, and most importantly for the prebiotic synthesis of adenine and guanine.

An important feature of all these energy sources is the activation of molecules in a local area followed by quenching of this activated mixture, and then protecting the organic compounds from further influence of the energy source. The quenching and protective steps are critical because the organic compounds will be destroyed if subjected continuously to the energy source.

B. Prebiotic Synthesis of Amino Acids

Mixtures of CH_4, NH_3, and H_2O with or without added H_2 are considered strongly reducing atmospheres. The atmosphere of Jupiter contains these species with the H_2 in large excess over the CH_4. The first successful prebiotic amino acid synthesis was carried out using this gas mixture and an electric discharge as an energy source.[6] The result was a large yield of amino acids (the yield of glycine alone was 2.1% based on the carbon), together with hydroxy acids, short aliphatic acids, and urea (Table II). One of the surprising results of this experiment was that the products were not a random mixture of organic compounds, but rather a relatively small number of compounds were produced in substantial yield. In addition the compounds produced were, with a few exceptions, of biological importance.

The mechanism of synthesis of the amino and hydroxy acids was investigated.[15] It was shown that the amino acids were not formed directly in the electric discharge but were the result of solution reactions of smaller

TABLE II.   Yields from Sparking a Mixture CH_4, NH_3, H_2O and H_2[a].

Compound	Yield (μmole)	Yield (%)
Glycine	630	2.1
Glycolic acid	560	1.9
Sarcosine	50	0.25
Alanine	340	1.7
Lactic acid	310	1.6
N-Methylalanine	10	0.07
α-Amino-n-butyric acid	50	0.34
α-Aminoisobutyric acid	1	0.007
α-Hydroxybutyric acid	50	0.34
β-Alanine	150	0.76
Succinic acid	40	0.27
Asparctic acid	4	0.024
Glutamic acid	6	0.051
Iminodiacetic acid	55	0.37
Iminoacetic-propionic acid	15	0.13
Formic acid	2330	4.0
Acetic acid	150	0.51
Propionic acid	130	0.66
Urea	20	0.034
N-Methyl urea	15	0.051
Total		15.2

[a] 59 mmoles (710 mg) of carbon were added as CH_4. The percent yields are based on the carbon.

molecules produced in the discharge, in particular hydrogen cyanide and aldehydes. The reactions are

$$RCHO + HCN + NH_3 \rightleftarrows RCH(NH_2)CN \xrightarrow{H_2O} RCH(NH_2)\overset{\overset{\displaystyle O}{\|}}{C}-NH_2$$

$$\xrightarrow{H_2O} RCH(NH_2)COOH$$

$$RCHO + HCN \rightleftarrows RCH(OH)CN \xrightarrow{H_2O} RCH(OH)\overset{\overset{\displaystyle O}{\|}}{C}-NH_2$$

$$\xrightarrow{H_2O} RCH(OH)COOH$$

These reactions were studied subsequently in detail, and the equilibrium and rate constants of these reactions were measured.[16] These results show that amino and hydroxy acids can be synthesized at high dilutions of HCN and aldehydes in a primitive ocean. It is also to be noted that the rates of these reactions were rather rapid. The half-lives for the hydrolysis of the amino and hydroxy nitriles are about 10^3 years.

This synthesis of amino acids, called the Strecker synthesis, requires the presence of NH_4^+ (and NH_3) in the primitive ocean. On the basis of the experimental equilibrium and rate constants it can be shown[16] that equal amounts of amino and hydroxy acids are obtained when the NH_4^+ concentration is about 0.01 M at pH 8 and 25°C with this NH_4^+ concentration being insensitive to temperature and pH. This translates into a pNH_3 in the atmosphere of 2×10^{-7} atm at 0° and 4×10^{-6} atm at 25°C. This is a low partial pressure, but it would seem to be necessary for amino acid synthesis. Ammonia is decomposed by ultraviolet light, but mechanisms for resynthesis are available. The details of the ammonia balance on the primitive earth remain to be worked out.

In a typical electric discharge experiment, the partial pressure of CH_4 is 0.1 to 0.2 atm. This pressure is used for convenience, and it is likely, but never demonstrated, that organic compound synthesis would work at much lower partial pressures of methane. There are no estimates available for pCH_4 on the primitive earth but 10^{-5} to 10^{-3} atm seems reasonable. Higher pressures are not reasonable because the sources of energy would convert the CH_4 to organic compounds in the oceans too rapidly for higher pressures of CH_4 to build up.

Ultraviolet light acting on this mixture of gases is not effective in producing amino acids except at very short wavelengths (< 1500 Å) and even then the yields are very low.[13] The low yields are probably due to the low yields of HCN produced by ultraviolet light. If the gas mixture is modified by adding gases such as H_2S or formaldehyde, then reasonable yields of amino acids can be obtained at relatively long wavelengths (< 2500 Å) where considerable energies from the sun are available.[14] The H_2S absorbs at these longer wavelengths and is photodissociated to H and HS. The H atoms have a high velocity and break up the CH_4 and NH_3. It is possible, but not demonstrated, that HCN and other molecules are produced, which then form amino acids in the aqueous part of the system.

Pyrolysis of CH_4 and NH_3 gives very low yields of amino acids. The pyrolysis conditions are from 800° to 1200°C with contact times of a second or less. However, the pyrolysis of CH_4 and other hydrocarbons gives good yields of benzene, phenylacetylene, and many other hydrocarbons. It can be shown that phenylacetylene would be converted to phenylalanine and tyrosine in the primitive ocean.[17] Pyrolysis of the hydrocarbons in the presence of NH_3 gives substantial yields of indole, which can be converted to tryptophan in the primitive ocean.

A mixture of CH_4, N_2, and traces of NH_3 and H_2O is a more realistic atmosphere for the primitive earth because large amounts of NH_3 would

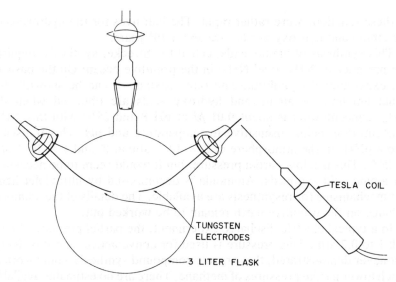

Fig. 1.   Electric discharge apparatus used to synthesize amino acids at room temperature.

not have accumulated in the atmosphere because the NH_3 would dissolve in the ocean. It is still, however, a strongly reducing atmosphere.

This mixture of gases is quite effective with an electric discharge in producing amino acids.[18] The yields are somewhat lower than with NH_3,

TABLE III.   Yields from Sparking CH_4 (336 mM), N_2, and H_2O of Traces of NH_3

Compound	Yield (μmole)	Compound	Yield (μmole)
Glycine	440	α,γ-Diaminobutyric acid	33
Alanine	790	α-Hydroxy-γ-aminobutyric acid	74
α-Amino-n-butyric acid	270	α,β-Diaminopropionic	6.4
α-Aminoisobutyric acid	~30	Isoserine	5.5
Valine	19.5	Sarcosine	55
Norvaline	61	N-Ethylglycine	30
Isovaline	~5	N-Propylglycine	~2
Leucine	11.3	N-Isopropylglycine	~2
Isoleucine	4.8	N-Methylalanine	~15
Alloisoleucine	5.1	N-Ethylalanine	<0.2
Norleucine	6.0	β-Alanine	18.8
$tert$-Leucine	<0.02	β-Anino-n-butyric acid	~0.3
Proline	1.5	β-Amino-isobutyric acid	~0.3
Aspartic acid	34	γ-Aminobutyric acid	2.4
Glutamic acid	7.7	N-Methyl-β-alanine	~5
Serine	5.0	N-Ethyl-β-alanine	~2
Threonine	~0.8	Pipecolic acid	~0.05
Allothreonine	~0.8		

but the products are more diverse (Table III). Hydroxy acids, short aliphatic acids, and dicarboxylic acids are produced along with the amino acids. Ten of the 20 amino acids that occur in proteins are produced directly in this experiment. Counting asparagine and glutamine, which are formed but hydrolyzed before analysis, and methionine, which is formed when H_2S is added,[19] one can say that 13 of the 20 amino acids in proteins can be formed in this single experiment. Cysteine was found in the photolysis of CH_4, NH_3, H_2O, and H_2S.[14] The pyrolysis of hydrocarbons, as discussed above, leads to phenylalanine, tyrosine, and tryptophan.[17,20] This leaves only the basic amino acids: lysine, arginine, and histidine. There are so far no established prebiotic syntheses of these amino acids. There is no fundamental reason that the basic amino acids cannot be synthesized, and this problem may be solved before too long.

C. Mildly Reducing and Nonreducing Atmospheres

There has been less experimental work with gas mixtures containing CO and CO_2 as carbon sources instead of CH_4, and the results so far have lower yields and a smaller variety of compounds.[21,22]

Electric discharges acting on a mixture of CO, N_2, and H_2 are not effective in amino acid synthesis unless the ratio of H_2 to CO is greater than about 1.0. Glycine is produced in fair yield, but only small amounts of any higher amino acids are produced. Large amounts of formaldehyde are obtained, however, and formaldehyde is important in the prebiotic synthesis of sugars.

A mixture of CO_2, N_2, and H_2 is more oxidized than the CO mixture, but the excess H_2 makes it a reduced mixture. As with CO + H_2, the amino acid synthesis is quite low with electric discharges unless H_2/CO_2 is greater than about 2. In this case glycine is produced in fair yield, but again very little of the higher amino acids are formed.

A mixture of CO + H_2 is used in the Fischer–Tropsch reaction to make hydrocarbons in high yields. The reaction requires a catalyst, usually Fe or Ni supported on silica, a temperatue of 200–400°C and a short contact time. Depending on the conditions, aliphatic hydrocarbons, aromatic hydrocarbons, alcohols, and acids can be produced. If NH_3 is added to the CO + H_2, then amino acids, purines, and pyrimidines can be formed.[23] The intermediates in these reactions are not known, but it is likely that HCN is involved together with some of the intermediates postulated for the electric discharge processes.

A mixture of CO + H_2O with electric discharges is not particularly effective in organic compound synthesis, but ultraviolet light that is absorbed by the water (< 1849 Å) results in the production of formaldehyde and other aldehydes, alcohols, and acids in fair yields.[24] The mechanism seems to involve splitting the H_2O to $H + OH$ with the OH converting CO to CO_2 and the H reducing another molecule of CO.

Electric discharges and ultraviolet light give no organic compounds with a mixture of $CO_2 + H_2O$. Ionizing radition (e.g., 40 MeV helium ions) gives small yields of formic acid and formaldehyde.[25]

The action of γ-rays on an aqueous solution of CO_2 and ferrous ion gives fair yields of formic acid, oxalic acid, and other simple products.[26] Ultraviolet light gives similar results. In these reactions, the Fe^{2+} is a stoichiometric reducing agent rather than a catalyst. Nitrogen in the form of N_2 does not react, and experiments with NH_3 have not been tried.

D. Purine and Pyrimidine Synthesis

Hydrogen cyanide is used in the synthesis of purines as well as amino acids. This is illustrated in a remarkable synthesis of adenine. If strong ammoniacal solutions of hydrogen cyanide are refluxed for a few days, adenine is obtained in up to 0.5% yield along with 4-aminoimidazole-5-carboxamide and the usual cyanide polymer.[27]

The mechanism of adenine synthesis in these experiments is probably

$$HCN + CN^- \rightarrow HN{=}HC{-}CN \xrightarrow{HCN} H_2N{-}CH(CN)_2$$

The difficult step in the synthesis of adenine just described is the reaction of tetramer with formadine. This step may be bypassed by the photochemical rearrangement of tetramer to aminoimidazole carboxamide, a reaction that proceeds readily in contemporary sunlight.[28]

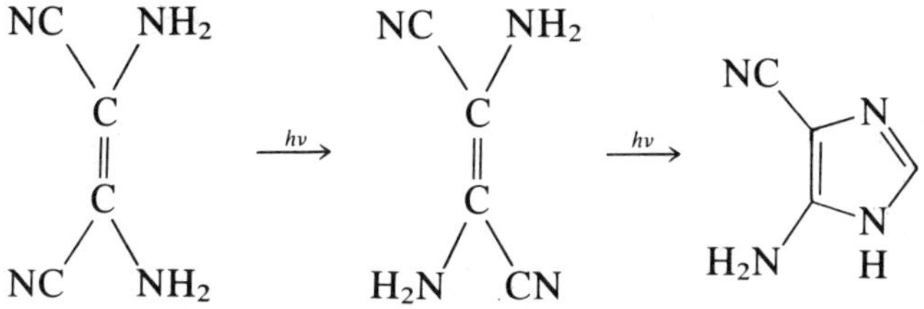

A further possibility is that tetramer formation may have occurred in an eutectic solution. High yield of tetramer ($> 10\%$) can be obtained by cooling dilute cyanide solutions to between $-10°C$ and $-30°C$ for a few months.

The prebiotic synthesis of the pyrimidine cytosine involves cyanoacetylene, which is synthesized in good yield by sparking mixtures of CH_4 + N_2. Cyanoacetylene reacts with cyanate to give cytosine,[29]

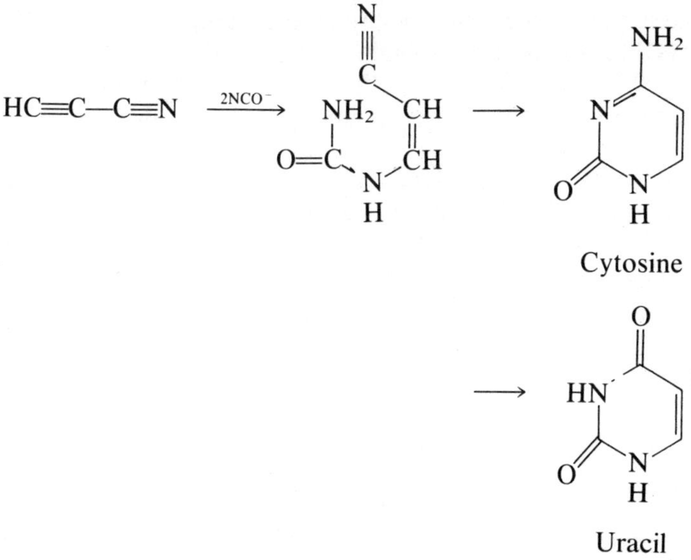

Cytosine

Uracil

and the cytosine can be converted to uracil. Cyanate can come from cyanogen or by the decomposition of urea.

Another prebiotic synthesis of uracil starts from β-alanine and cyanate and ultraviolet light.[30]

$$H_2N—CH_2—CH_2—COOH \xrightarrow{NCO^-} \quad O{=}C \quad CH_2 \longrightarrow$$

(structure with H₂N and COOH over the central carbons, HN—CH₂ below)

(intermediate ring structure with HN, CH₂, CH₂, C=O, N—H)

$$\xrightarrow{h\nu} \quad$$ (uracil ring structure: HN, C=O, O, N, H)

E. Sugars

The synthesis of reducing sugars from formaldehyde under alkaline conditions was discovered long ago. However, the process is very complex and incompletely understood. In simple solution, it depends on the presence of a suitable catalyst. Calcium hydroxide and calcium carbonate are among the most popular heterogeneous catalysts. In the absence of catalysts, little or no sugar is obtained. Particularly attractive is the finding that at 100°C, clays such as kaolin serve to catalyze formation of monosaccharides, including ribose, in good yield from dilute (0.01 M) solutions of formaldehyde.[31,32]

The reaction is autocatalytic and proceeds in stages through glycolaldehyde, glyceraldehyde, and dihydroxyacetone, tetroses, and pentoses to give finally hexoses including glucose and fructose. One proposed reaction sequence is

$$CH_2O \rightarrow \begin{array}{c} CHO \\ | \\ CH_2OH \end{array} \rightarrow \begin{array}{c} CHO \\ | \\ CHOH \\ | \\ CH_2OH \end{array} \leftrightarrows \begin{array}{c} CH_2OH \\ | \\ C{=}O \\ | \\ CH_2OH \end{array} \rightarrow \begin{array}{c} CH_2OH \\ | \\ C{=}O \\ | \\ CHOH \\ | \\ CH_2OH \end{array} \rightarrow \begin{array}{c} CHO \\ | \\ CHOH \\ | \\ CHOH \\ | \\ CH_2OH \end{array}$$

―――――― reverse aldol ――――――

$$\rightarrow pentoses \rightarrow hexoses$$

The problem with sugars on the primitive earth is not their synthesis, but rather their stability. They decompose in a few hundred years at most at 25°C. There are a number of possible ways to stabilize sugars, the most interesting being to convert the sugar to a glycoside of a purine or pyrimidine.

F. Organic Compounds in Carbonaceous Chondrites

On 28 September 1969 a type II carbonaceous chondrite fell in Murchison, Australia. Surprisingly large amounts of amino acids were found by Kvenvolden et al.[33,34] The first report identified seven amino acids (glycine, alanine, valine, proline, glutamic acid, sarcosine, and α-aminoisobutyric acid), of which all but valine and proline had been found in the original electric discharge experiments.[6,15] The most striking are sarcosine and α-aminoisobutyric acid. The second report identified 18 amino acids of which nine had previously been identified in the original electric discharge experiments, but the remaining nine had not.

At that time we had identified the hydrophobic amino acids from the low temperature electric discharge experiments described above, and therefore we examined the products for the nonprotein amino acids found in Murchison. We were able to find all of them.[18]

There is a striking similarity between the products and relative abundances of the amino acids produced by electric discharge and the meteorite amino acids. Table IV compares the results. The most notable difference between the meteorite and the electric discharge amino acids is the pipecolic acid, the yield being extremely low in the electric discharge. Proline is also present in relatively low yield from the electric discharge. The amount of α-aminoisobutyric acid is greater than α-amino-*n*-butyric acid in the meteorite, but the reverse is the case in the electric discharge. We do not believe that reasonable differences in ratios of amino acids detract from the overall picture. Indeed, the ratio of α-aminoisobutyric acid to glycine is quite different in two meteorites of the same type, being 0.4 in Murchison and 3.8 in Murray.[35] A similar comparison has been made between the dicarboxylic acids in Murchison[36] and those produced by an electric discharge,[37] and the product ratios are quite similar.

The close correspondence between the amino acids found in the Murchison meteorite and those produced by an electric discharge synthesis, both as to the amino acids produced and their relative ratios, suggests that the amino acids in the meteorite were synthesized on the parent body by means of an electric discharge or analogous processes. Electric discharges appear to be the most favored source of energy but sufficient data are not available to make realistic comparison with other energy sources. In any case, it is unlikely that a single source of energy synthesized all of the organic compounds either on the parent body of the carbonaceous chondrites or on the primitive earth. All sources of energy

TABLE IV. Relative Abundances of Amino Acids in the Murchison Meteorite and in
an Electric Discharge Synthesis[a]

Amino acid	Murchison meteorite	Electric discharge
Glycine	****	****
Alanine	****	****
α-Amino-n-butyric acid	***	****
α-Aminoisobutyric acid	****	**
Valine	***	**
Norvaline	***	***
Isovaline	**	**
Proline	***	*
Pipecolic acid	*	<*
Aspartic acid	***	***
Glutamic acid	***	**
β-Alanine	**	**
β-Amino-n-butyric acid	*	*
β-Aminoisobutyric acid	*	*
γ-Aminobutyric acid	*	**
Sarcosine	**	***
N-Ethylglycine	**	***
N-Methylalanine	**	**

[a] Mole ratio to glycine (100): 0.05–0.5, *; 0.5–5, **; 5–50, ***; >50 ****.

would have made their contribution, and the problem is to evaluate the relative importance of each source.

Our ideas on the prebiotic synthesis of organic compounds are based largely on the results of experiments in model systems. So it is extremely gratifying to see that such synthesis really did take place on the parent body of the meteorite, and so it becomes quite plausible that they took place on the primitive earth.

III. INTERSTELLAR MOLECULES

In the past 10 years a large number of organic molecules have been found in interstellar dust clouds mostly by emission lines in the microwave region of the spectrum (for a summary see Ref. 38). The concentration of these molecules is very low (a few molecules per cm^3 at the most) but the total amount in a dust cloud is large. The molecules found include formaldehyde, hydrogen cyanide, acetaldehyde, and cyanoacetylene. These are important prebiotic molecules, and this immediately raises the question of whether the interstellar molecules played a role in the origin of life on the earth. In order for this to have taken place it would have been necessary for the molecules to have been greatly concentrated in the solar nebula and to have arrived on the earth without being destroyed by ultraviolet light or pyrolysis. This appears to be difficult to do. In addition it is necessary for some molecules to be continuously synthesized (unless life started very quickly) because of their instability, and an interstellar source could not be responsible for these.

For these reasons, it is generally felt that the interstellar molecules played at most a minor role in the origin of life. However, the presence of so many molecules of prebiotic importance in interstellar space, combined with the fact that their synthesis must differ from that on the primitive earth where the conditions were very different, indicates that some molecules are particularly easily synthesized when radicals and ions recombine. Another way of saying this is that there appears to be a universal organic chemistry, which shows up in interstellar space, in the atmospheres of the major planets, and in the reducing atmosphere of the primitive earth.

IV. POLYMERIZATION PROCESSES

All prebiotic polymerization reactions, which are dehydration reactions, are thermodynamically unfavorable. This free energy barrier can be overcome in two ways. The first is to drive the dehydration reaction by coupling it to the hydration of a high energy compound, and the second method is to remove the water by heating. In principle, visible or ultraviolet light could drive these reactions, but so far no one has demonstrated adequately such processes.

A. Peptide Synthesis

An example of the use of a high energy compound to synthesize a peptide is cyanamide, which is hydrated to urea.

$$2 \ H_3N^+—CH_2—COO^- \ + \ H_2N—C{\equiv}N{\rightarrow}$$

$$H_3N^+—CH_2—\overset{\overset{\displaystyle O}{\|}}{C}—NH—CH_2—COO^- \ + \ H_2N—\overset{\overset{\displaystyle O}{\|}}{C}—NH_2$$

The yields of this reaction are at best a few percent.[39] Similar results are obtained with cyanate, cyanoguanidine, cyanogen and the tetramer of HCN (for review, see Ref. 40). One of the problems with these reagents is that they tend to react with the amino group [to give the glycocyamine (I) in the case of cyanamide] rather than activating the carboxyl group

(II)

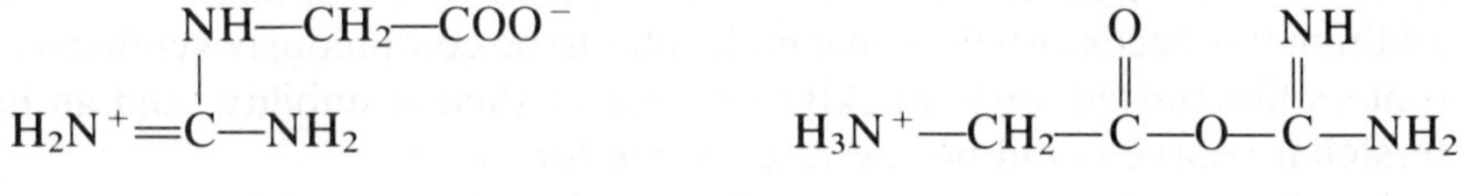

$$\begin{array}{c} NH{-}CH_2{-}COO^- \\ | \\ H_2N^+{=}C{-}NH_2 \end{array} \qquad\qquad H_3N^+{-}CH_2{-}\overset{\overset{\displaystyle O}{\|}}{C}{-}O{-}\overset{\overset{\displaystyle NH}{\|}}{C}{-}NH_2$$

(I)   (II)

for subsequent nucleophilic attack by the amino group of a second amino acid. Polyphosphates and ATP give somewhat higher yields of dipeptides, but these phosphates also phosphorylate the nitrogen which can block peptide synthesis.

One way around this problem is to use imidazole as a catalyst.[41,42] The ATP reacts with the amino group of the amino acid forming the phosphoroamidate which does not polymerize. In the presence of imidazole (Im) the phosphoroamidate of the amino acid is converted to the imidazolide (ImPA).

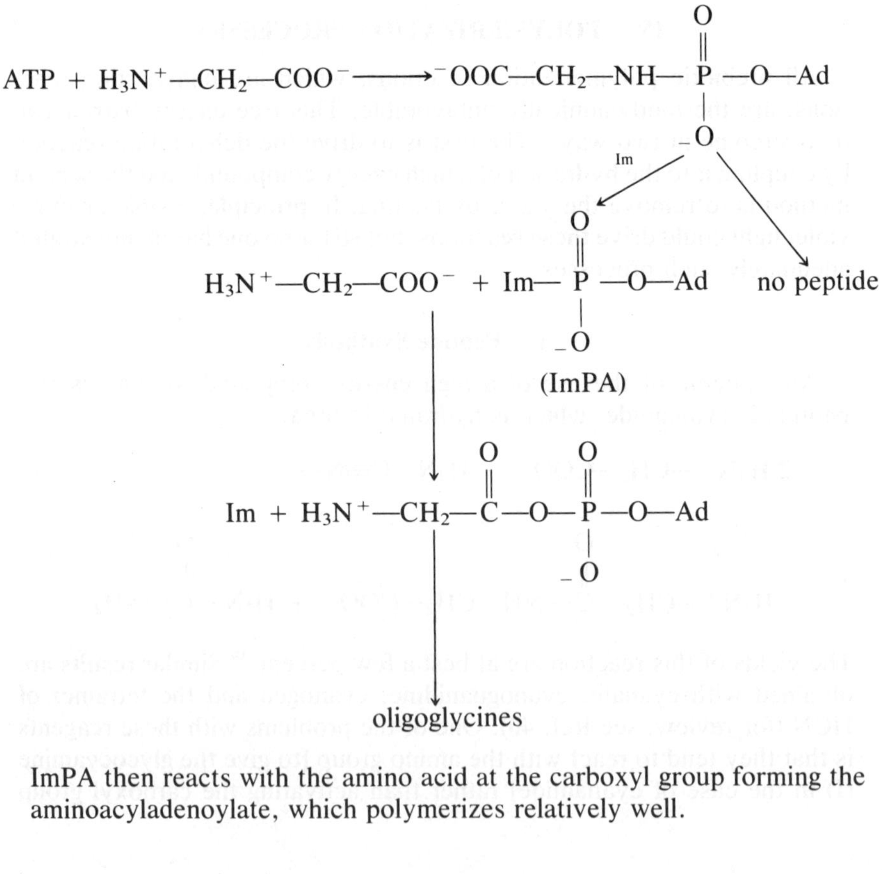

$$ATP + H_3N^+{-}CH_2{-}COO^- \longrightarrow {}^-OOC{-}CH_2{-}NH{-}\overset{\overset{\displaystyle O}{\|}}{\underset{\underset{\displaystyle {}_-O}{|}}{P}}{-}O{-}Ad$$

$$H_3N^+{-}CH_2{-}COO^- + Im{-}\overset{\overset{\displaystyle O}{\|}}{\underset{\underset{\displaystyle {}_-O}{|}}{P}}{-}O{-}Ad \qquad \text{no peptide}$$

(ImPA)

$$Im + H_3N^+{-}CH_2{-}\overset{\overset{\displaystyle O}{\|}}{C}{-}O{-}\overset{\overset{\displaystyle O}{\|}}{\underset{\underset{\displaystyle {}_-O}{|}}{P}}{-}O{-}Ad$$

oligoglycines

ImPA then reacts with the amino acid at the carboxyl group forming the aminoacyladenoylate, which polymerizes relatively well.

The other approach is to heat mixtures of amino acids at low relative humidities. This process has been investigated extensively by Fox and co-workers.[43] Typical conditions involve using an excess of aspartic and glutamic acids and heating to 150–180°C for a few hours. The yields are rather good (~ 50%) and molecular weights of a few thousand can be obtained. The problem with this reaction is that dry temperatures of 150–180°C do not occur extensively on the earth. Higher temperatures are available in volcanoes (1200°C), but this would destroy the amino acids. Temperatures of 150–180°C can be obtained at sufficient depth in the earth's crust, but long heating times would also destroy the amino acids. Hot springs are not suitable because of the excess water.

A few experiments with these thermal polymerizations have been done at lower temperatures[44] but the yields are small, and it is not clear whether the results obtained at 150–180° can be obtained over longer periods at the more reasonable temperatures of 50–100°C.

B.   Prebiotic Synthesis of Nucleosides and Polynucleotides

One of the more difficult prebiotic syntheses is that of the nucleosides. Heating of ribose and purines at 100°C gives fair yields (2–20%) of a mixture of isomers of the purine ribosides, with some of the correct β-isomer being produced.[45] However, a mixture of pyrimidines and ribose gives no detectable yield of nucleosides. Whether the inability to synthesize pyrimidine ribosides is a matter of not trying the correct conditions or whether pyrimidine nucleosides were not involved in the first organisms remains to be determined.

The phosphorylation of nucleosides presents less problems. The first step is the synthesis of polyphosphates from ammonium dihydrogen phosphate. Dihydrogen phosphates do not occur on the earth at the present time, and even monohydrogen phosphates are rare. The reason for this is that hydroxyl apatite $[Ca_{10}(PO_4)_6(OH)_2]$ is the major phosphate mineral, and it is very insoluble. (The dissolved phosphate in equilibrium with apatite in the oceans is $3 \times 10^{-6} M$.) However, the precipitation of apatite is slow especially in the presence of Mg^{2+}.[46] Phosphate does not build up in the ocean at the present time because organisms remove it. However, on the primitive earth there might be a build up of phosphate, and the presence of NH_4^+ in the oceans would aid this. One can then envision the evaporation of sea water in a lagoon and the precipitation of Mg-NH_4PO_4 and $NH_4H_2PO_4$. On heating to 80–100°, especially in the presence of urea, polyphosphates are produced in good yield.[47]

These polyphosphates can then phosphorylate nucleosides rather efficiently.[48] If amino acids are present then AMP-amino acids are synthesized (see under peptide synthesis), but if imidazole is present then im-

Adenosine + polyphosphates $\xrightarrow{\text{solution}}$ adenosine polyphosphates

Adenosine + polyphosphates + amine $\xrightarrow[\substack{Mg^{2+} \\ 37-100°C}]{\text{dry out}}$ AMP − NHR

Adenosine + polyphosphates + imidazole $\xrightarrow[\substack{Mg^{2+} \\ 37-100°C}]{\text{dry out}}$ AMP − imidazolide (ImPA)

AMP − imidazolide $\xrightarrow[\text{template}]{\text{solution}}$ template polymerization

Fig. 2.   Prebiotic synthesis of adenosine polyphosphates, adenosine phosphoroimidates and the imidazolide of AMP.

idazolides are produced, for example, ImPA. The imidazolides are very reactive compounds which can be used for template polymerizations. These reactions are summarized in Fig. 2.

Imidazole is a reasonable prebiotic compound, since it can be synthesized from glyoxal + $2NH_3$ + H_2CO or from cyanoacetylene + NH_3 + ultraviolet light. However, it is not clear whether large amounts were present in the primitive ocean. It may be that the template polymerizations described below were carried out slowly using nucleotide polyphosphates. Such experiments are difficult to carry out in the lab because of the long time needed.

The template polymerizations carried out by Orgel and co-workers[49,50] is the most exciting area of prebiotic chemistry. If such polymerizations could be done with adequate fidelity then the basis for the development of genetic information would be available. The results of an extensive series of experiments are summarized in Fig. 3.

The first result is that pyrimidine nucleosides do not give template polymerizations under the conditions tried. This is attributed to the inability of pyrimidine nucleosides to stack on the poly A or poly G template. It is possible that early template polymerizations used dimers, trimers, etc. which may stack, or that there may be conditions not yet found which

Poly U + ImP A $\xrightarrow{Pb^{2+}\ or\ Zn^{2+}}$ Poly A (2′,5′)

Poly C + ImP G $\xrightarrow{Pb^{2+}\ or\ Zn^{2+}}$ Poly G $\substack{(2′,5′ \text{ with } Pb^{2+}) \\ (3′,5′ \text{ with } Zn^{2+})}$

Poly A + ImP U $\xrightarrow{\hspace{2cm}}$ no template polymerization

Poly G + ImP C $\xrightarrow{\hspace{2cm}}$ no template polymerization

Fig. 3.   Results of template polymerizations of different nucleotide imidazolides.

permit the stacking of pyrimidine nucleosides. Otherwise it is difficult to see how pyrimidines were involved in the early genetic system.

The second result is that $2'-5'$ phosphodiester bonds are usually formed more easily than the desired $3'-5'$ bonds. The $2'-5'$ bonds hydrolyze more readily than $3'-5'$ bonds which would shift the population of polynucleotides in the primitive ocean over the $3'-5'$, but it clearly would be better to obtain the $3'-5'$ isomers. This can be done by a Zn^{2+} catalyst as long as the Zn^{2+} is not complexed with the buffer salts.[50]

The most interesting aspect of the Zn^{2+}-catalyzed polymerization is its fidelity. If equal amounts of ImPA and ImPG are reacted with a poly C template, only 1 A per 200 G is incorporated into the poly G. This is only an order of magnitude less fidelity than RNA polymerases (about 1 in 4000).

C. Future Work

The above review shows the progress that has been made in the last 30 years. The prebiotic synthesis of amino acids, purines, pyrimidines, and sugars is understood at a basic level, although more details of the reactions are needed. The polymerization processes are less well understood, and while some of them are plausible it is necessary to work them out in greater detail. The template polymerization reactions are an exciting beginning and may show how genetic information started to accumulate. So far the problem of nucleic acid directed enzyme synthesis has not been dealt with on an experimental level. The problems in this area, which are very difficult, are considered by other speakers in this symposium.

Acknowledgment

This work has been supported by NASA Grant No. NAGW-20.

References

1. A. I. Oparin, *The Origin of Life*, Macmillan, New York, 1938.
2. J. B. S. Haldane, *Rationalist Annual*, **148**, 3 (1929); reprinted in *Science and Human Life*, Harper Bros., New York and London, 1933, p. 149.
3. H. C. Urey, *Proc. Natl. Acad. Sci. USA* **38**, 351–363 (1952); also in *The Planets*, Yale University Press, New Haven, Connecticut, pp. 149–157.
4. J. D. Bernal, *The Physical Basis of Life*, Routledge and Kegan Paul, London, 1951.
5. N. H. Horowitz, *Proc. Natl. Acad. Sci. USA*, **31**, 153–157 (1945).
6. S. L. Miller, *Science*, **117**, 528–529 (1953).
7. S. L. Miller and L. E. Orgel, *The Origins of Life on the Earth*, Prentice Hall, Englewood Cliffs, New Jersey, 1974.
8. D. H. Kenyon and G. Steinman, *Biochemical Predestination*, McGraw-Hill, New York, 1969.
9. R. M. Lemmon, *Chem. Rev.*, **70**, 95–109 (1970).

10. M. R. Walter, R. Buick, and J. S. R. Dunlop, *Nature (London)*, **284**, 443–445 (1980); S. M. Awramik, J. W. Schopf, and M. R. Walter, *Precambriam Res.* to be published (1983).

11. S. L. Miller, H. C. Urey, and J. Oró, *J. Mol. Evol.*, **9**, 59–72 (1976).

12. A. Bar-Nun, N. Bar-Nun, S. H. Bauer, and C. Sagan, *Science*, **168**, 470–473 (1970).

13. W. Groth and H. von Weyssenhoff, *Planet. Space Sci.*, **2**, 79–85 (1960).

14. C. Sagan and B. N. Khare, *Science*, **173**, 417–420 (1971); *Nature (London)*, **232**, 577–578 (1971).

15. S. L. Miller, *Ann. N.Y. Acad. Sci.*, **69**, 260–274; also in *The Origin of Life on the Earth*, A. Oparin, ed., Pergamon Press, Oxford, 1959, pp. 123–135.

16. J. E. Van Trump and S. L. Miller, in *Origin of Life*, Y. Wolman, ed., Reidel, Dordrecht, Holland, 1981, pp. 135–141.

17. N. Friedmann and S. L. Miller, *Science*, **166**, 766–767 (1969).

18. D. Ring, Y. Wolman, N. Friedmann, and S. L. Miller, *Proc. Natl. Acad. Sci. USA* **69**, 765–768 (1972); Y. Wolman, W. J. Haverland, and S. L. Miller, *Proc. Natl. Acad. Sci. U.S.A.*, **69**, 809–811 (1972).

19. S. L. Miller and J. E. Van Trump, *Science*, **178**, 859–860 (1972).

20. N. Friedmann, W. J. Haverland, and S. L. Miller, in *Chemical Evolution and the Origin of Life*, R. Buret and C. Ponnamperuma, eds. North Holland, Amsterdam, 1971, pp. 123–135.

21. P. H. Abelson, *Proc. Natl. Acad. Sci. U.S.A.*, **54**, 1490–1494 (1965).

22. G. Schlesinger and S. L. Miller, *J. Mol. Evol.* to be published (1983).

23. R. Hatyatsu et al., *Geochim. Cosmochim. Acta*, **36**, 555–571 (1972); D. Yoshino, R. Hayatsu, and E. Anders, *Geochim. Cosmochim. Acta*, **35**, 927–938 (1971).

24. A. Bar-Nun and H. Hartman, *Orig. Life*, **9**, 93–101 (1978).

25. W. M. Garrison, D. C. Morrison, J. G. Hamilton, A. A. Benson, and M. Calvin, *Science*, **114**, 416–418 (1951).

26. N. Getoff, *Z. Naturforsch*, **17b**, 87–90, 751–757 (1962).

27. J. Oró and A. P. Kimball, *Arch. Biochem. Biophys.*, **94**, 221–227 (1961); *ibid.* **96**, 293–313 (1962).

28. R. A. Sanchez, J. P. Ferris, and L. E. Orgel, *J. Mol. Biol.*, **30**, 223–253 (1967); *ibid.* **38**, 121–128 (1968).

29. R. A. Sanchez, J. P. Ferris, and L. E. Orgel, *Science*, **154**, 784–785 (1966); J. P. Ferris, R. A. Sanchez, and L. E. Orgel, *J. Mol. Biol.*, **33**, 693–704 (1968).

30. A. W. Schwartz and G. J. F. Chittenden, *Biosystems*, **9**, 87–92 (1977).

31. N. W. Gabel and C. Ponnamperuma, *Nature (London)*, **216**, 453–455 (1967).

32. C. Reid and L. E. Orgel, *Nature (London)*, **216**, 455 (1967).

33. K. Kvenvolden, J. G. Lawless, K. Pering, E. Peterson, J. Flores, C. Ponnamperuma, I. R. Kaplan, and C. Moore, *Nature (London)*, **228**, 923–926 (1970).

34. K. A. Kvenvolden, J. G. Lawless, and C. Ponnamperuma, *Proc. Natl. Acad. Sci. USA*, **68**, 486–490 (1971).

35. J. R. Cronin and C. B. Moore, *Science*, **172**, 1327–1329 (1971).

36. J. G. Lawless, B. Zeitman, W. E. Pereira, R. E. Summons, and A. M. Duffield, *Nature (London)*, **251**, 40–42 (1974).

37. B. Zeitman, S. Chang, and J. G. Lawless, *Nature (London)*, **251**, 42–43 (1974).

38. A. P. C. Mann and D. A. Williams, *Nature (London)*, **283**, 721–725 (1980).

39. C. Ponnamerpuma and E. Peterson, *Science,* **147,** 1572 (1965).

40. J. Hulshof and C. Ponnamperuma, *Orig. Life,* **7,** 197–224 (1976).

41. R. Lohrmann and L. E. Orgel, *Nature (London),* **224,** 418–420 (1973).

42. A. L. Weber and L. E. Orgel, *J. Mol. Evol.,* **11,** 9–16 (1978).

43. S. W. Fox and K. Dose, *Molecular Evolution and the Origin of Life,* Freeman, San Francisco, 1972.

44. D. L. Rohlfing, *Science,* **193,** 68–70 (1976).

45. W. D. Fuller, R. A. Sanchez, and L. E. Orgel, *J. Mol. Biol.,* **67,** 25–33 (1972); *J. Mol. Evol.,* **1,** 249–257 (1972).

46. G. J. Handschuh and L. E. Orgel, *Science,* **179,** 483 (1973).

47. R. Osterberg and L. E. Orgel, *J. Mol. Evol.,* **1,** 241–248 (1972).

48. R. Lohrmann, *J. Mol. Evol.,* **8,** 197–210 (1976); *J. Mol. Evol.,* **10,** 137–154 (1977).

49. R. Lohrmann and L. E. Orgel, *Nature (London),* **261,** 342–344 (1976).

50. R. Lohrmann, P. K. Bridson, and L. E. Orgel, *Science,* **208,** 1464–1465 (1980).

COMMENTARY: ON THE ORIGIN OF OPTICAL ACTIVITY

PETER SCHUSTER

Institut für Theoretische Chemie und Strahlenchemie
Universität Wien, Austria

At some stage in prebiotic chemistry surface catalysis became a condition sine qua non for an efficient synthesis of biopolymers. "Surface" should be understood in the widest sense: the surface may well be also the accessible part of some macromolecule. Powerful surface catalysis will make use of chirality or, in other words, if one antipode of a chiral molecule fits to an active center the opposite would not. Thus, we can imagine two synthetic systems being perfect mirror images. These two systems are not independent, since they compete for the achiral material required. Selection will enhance any existing difference in concentration of the two subsystems provided the rate of synthesis is autocatalytic of an order larger than 1. Ultimately, one system becomes extinct and its mirror image becomes established. The first difference between both systems, actually the origin of optical activity may be the result of a local fluctuation in an open system which is enhanced to the macroscopic level. Such fluctuations were observed experimentally. One way to visualize and verify this was reported by Havinga[1] and more recently by Thiemann[2]: slow crystallization of a racemic mixture of a chiral compound (which does not form racemic mixed crystals) sometimes yields an optically active solid and an optically active solution due to the fact that formation of a nucleus for crystallization is a rare event. Liquid and solid can be separated easily by some unspecific mechanical force. The whole complex of problems has been discussed also by Wald.[3]

References

1. E. Havinga, *Biochim. Biophys. Acta,* **13,** 171 (1954).
2. W. Thiemann, *J. Mol. Evol.,* **4,** 85 (1974).
3. G. Wald, *Ann. N.Y. Acad. Sci.,* **69,** 353 (1957).

COMMENTARY: ON SELECTION OF CHIRALITY

RICHARD M. NOYES

Department of Chemistry,
College of Liberal Arts,
University of Oregon, Eugene, Oregon

It is probably correct that a random genetic difference would have been sufficient to enable one of two chiral forms of living matter to become dominant. It might also have been that both chiral life forms developed in the primordeal ocean when it was rich in nutrient molecules. They need not initially have competed with each other. However, the metabolic waste products like carbon dioxide and methane are achiral. The form of life that developed photosynthesis could "recycle" those waste products into chiral material of use only to it and would then have had an overwhelming selective advantage.

COMMENTARY: A PREBIOTIC ORIGIN OF PHOTOSYNTHESIS?

C. SYBESMA

*Biophysics Laboratory, Vrije Universiteit Brussel,
Brussels, Belgium*

The usual consensus sets the origin of photosynthesis some $3-3.2 \times 10^9$ years back,[1] well within the biological era of evolution and *after* the first appearance of anaerobic bacteria. The oxygen evolving cyanobacteria would have emerged between 2.7 and 3×10^9 years ago, thus helping to set the conditions for the evolution of nonphotosynthetic oxygen requiring organisms. The stromatolites found in Southern Rhodesia and dated $(2.7-3) \times 10^9$ years back are the earliest fossil evidence of photosynthetic organisms (cyanobacteria). The generally supported assumption that the start of photosynthetic bacteria came after the development of anaerobic heterotrophs is based on the Darwinian hypothesis that when energy-rich nutricents for the anaerobes in the environment became depleted a selective advantage arose for those organisms that have learned to synthesize energy-rich compounds making use of sunlight.

In a recent publication, Folsome[2] proposed the idea that the polymerization necessary for "growth" and single-line descent of the prebiotically formed microstructures occurred by photochemical activation reactions. In this implicit suggestion of a prebiotic origin of photosynthesis, the activation of the "target molecules" (the "Miller–Urey compounds"[3]), to form polymers, is thought to take place with "high-energy" phosphates, a process that is essentially similar to what occurs in present day life. The high-energy phosphates would have been formed in a photochemical reaction in which a photoinduced charge separation within the boundary of the microsphere results in a proton translocation (perhaps a rearrangement of hydrogen bonds). This proton translocation (relocation) then takes care of phosphorylation reactions much in the same way as it occurs presently. The photoreceptors would have been porphyrin-like molecules formed in the environment and adsorbed to the boundary of the microsphere. Indeed, porphyrins could have been found under the conditions of the prebiotic earth, although firm evidence is still lacking.[4]

Folsome's idea can be seen as an argument for setting the origin of photosynthesis at a much earlier stage, in fact, in the prebiotic era. The purpose of this intervention is to present some evidence that may support such a prebiotic origin of photosynthesis.

Higher plant photosynthesis as it occurs today is a well-advanced form of the process. Cyanobacteria carry out the process much in the same way as higher plants do (although less is known about the details of cyanobacterial photosynthesis than is about the process in chloroplasts). In this process the energy is captured in the form of reducing power as well as in the form of high-energy phosphates (adenosine triphosphate, ATP). Photosynthesis in photosynthetic bacteria is more primitive and seems to be aimed primarely to the production of high-energy phosphates. The essential feature of the process in plants as well as in bacteria, however, is the transport of electronic charge in the photosynthetic membrane, coupled to a translocation (or relocation) of protons across or within the membrane (Fig. 1). The translocated (relocated) protons represent an electrochemical potential that is used to add the terminal phosphate to ADP, thus synthesizing ATP. This reaction is catalyzed by an enzyme complex, the so called coupling factor (CF). The CF can be made to catalyze the reverse reaction, thus acting like an ATPase and causing a reversed flow of protons. The electronic charge separation occurs in another protein complex (the reaction center). This protein from photosynthetic bacteria has been isolated and studied in detail.[5] It is composed of three subunits and binds chlorophyll, ubiquinone, and iron. Further (secondary) electron transport mediating proteins are cytochromes and ferredoxin, the latter being a nonheme iron–sulfur protein, found in several different kinds of organisms and mediating electron transport at a low redox potential.

Evolutionary relationships of electron carrying proteins from different organisms have been established by analysis and comparison of amino acid sequences. This was done with cytochrome c[6,7] and with ferredoxin.[8,9] Baltscheffsky[10] analyzed evolutionary relationships of three-dimensional structures. He proposed a line of descent in which all present-day electron carriers evolved from a ferredoxin-like ancestral protein operating at a low redox potential level in primitive electron transport reactions. The ancestral protein would have had an antiparallel β-structure. According to Orgel[11] and Carter and Kraut,[12] antiparallel β-structures may have been of prebiotic origin, functioning as very early and essential components in evolving polypeptides. Von Heyne et al.[13] showed that the apoprotein of ferredoxin has a strong β-structure forming capability. The fact that the β-structure is not realized in the final configuration of

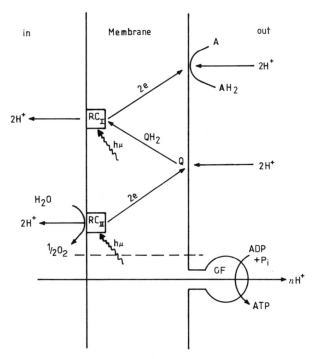

Fig. 1. Schematic representation of the chemiosmotic mechanism of photosynthetic energy conversion in higher plants. Two light reactions, each causing an electronic charge separation (2e) operate in series. In RC II a strong oxidant is produced which oxidizes water, leaving two protons in the inner medium. The electrons reduce an intermediate Q which binds two protons from the outer medium. QH_2 then re-reduces RC I, leaving the protons in the inner medium. The electronic charge from RC I causes the reduction of the terminal acceptor A, again with the addition of two protons from the outer medium. The protons, thus, generate a gradient across the membrane. This proton gradient "discharges" through the coupling factor CF with the concomitant phosphorylation of ADP to ATP.

the protein may be due to its binding to structures like the iron–sulphur clusters found in ferredoxin.

One of the conclusions of this study was that the binding of the metal–sulphur clusters would have caused a rearrangement of hydrogen bond patterns. This rearrangement was suggested to be, not just linked to the mechanism (of electron transport) but part of the mechanism itself.[14] Such a rearrangement of hydrogen bonds may be part of the coupling mechanism of the phosphorylation coupled to electron transport, at least in the more primitive and tightly membrane bound coupling enzyme which was found in the photosynthetic bacterium *Rhodospirillum rubrum*[15] and recently also in *Chromatium vinosum*.[16] This primitive coupling factor couples electron transport (or, rather, proton translocation) to the phosphorylation of inorganic phosphate to pyrophosphate, a reaction that is insensitive to well-known inhibitors of the ATP phosphorylation. The factor may very well be an evolutionary precursor of the CF-ATPase present in the more developed organisms.

Some evidence for a much closer functional relationship between the electron transport and proton translocation on one hand and the phosphorylation on the other hand in chloroplasts is given by a recent report from Prochaska and Dilley.[17] This report shows that in chloroplasts the protons released from water in the photoreaction of photosystem II are sequestered in the membrane and interact directly with a particular subunit of the CF *within* the membrane. Recent evidence from other laboratories, including our own,[18,19] all points to a much more localized mech-

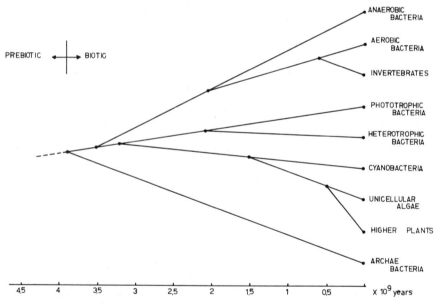

Fig. 2. An evolution diagram illustrating a suggestion of common ancestry of some present-day organisms. The essential features of present-day photosynthesis may have originated in the prebiotic era and is preserved in its most primitive form in (at least some) present-day phototrophs. The heterotrophs may have developed parallel with the aerobic nonphotosynthetic bacteria, some 1 to 1.5×10^9 years after the emergence of the cyanobacteria. The eukaryotic photosynthetic organisms developed much later, perhaps some 1.5 to 0.5 $\times 10^9$ years ago. The archaebacteria are primitive organisms that seem to have no evolutionary relation with the present prokaryotes.[21] Little is known about their energy metabolism. Tentatively, they are considered as a very early form of cellular life.

anism of the coupling than would be suggested by Mitchell's original hypothesis.[20]

It does not seem unreasonable to suggest that the essential features of the coupling mechanism have been preserved during the course of evolution and, if protocells "grew" and formed single lines of descent by a photochemical mechanism, that from this mechanism photosynthesis evolved (Fig. 2). Thus, one could think of a following sequence of events:

1. In the prebiotic period microspheres have been formed by self-assembly of polymeric material.[4] Random peptides, which are bound or adsorbed to the boundary of these microsystems, would adopt preferentially a β-structure.
2. The β-structures easily could have bound metals (iron), thus making them suitable for electron transport.
3. If, on the boundary, also primitive photoreceptors (porphyrin) were adsorbed, excitation by the radiation from the sun could have led to electron displacement in the polypeptides, causing changes in hydrogen bond patterns.
4. The changes of hydrogen bond patterns could result in the production of pyrophosphate (and, perhaps also organic phosphates).
5. Increase of the polymeric content of the microspheres then could occur through activation, by phosphate bonds, of the monomers.
6. Thus, the microspheres could "grow" and spontaneous fission could have formed single lines of descent.

Some plausibility for such a "scenario" could be inferred from the following:

1. β-structures are primitive structures that are preserved in many electron transport proteins (or apoproteins at least), phosphorylation coupling factors and chlorophyll–protein complexes.[22]
2. β-structures contain hydrogen bond patterns that may be changed or broken upon interaction with atoms (metals), clusters, and molecular groups (nucleotides).
3. Phosphorylation occurs by interaction with hydrogen ions, probably directly at the level of particular amino acids or subunits of the proteins mediating the coupling.

References

1. J. W. Schöpf, *Sci. Am.*, **293** No. 3, 84 (1978).
2. C. E. Folsome, *The Origin of Life*, Freeman, San Francisco, 1979.
3. S. L. Miller and H. C. Urey, *Science*, **130**, 245 (1959).

4. S. W. Fox and K. Dose, *Molecular Evolution and the Origin of Life*, Freeman, San Francisco, 1972.

5. R. C. Prince and P. L. Dutton, in *The Photosynthetic Bacteria*, R. K. Clayton and W. R. Sistrom, eds., Plenum Press, New York, 1978, pp. 439–453.

6. W. M. Fitch and E. Margoliash, *Science*, **155**, 279 (1967).

7. G. W. Moore, M. Goodman, C. Callahan, R. Molmquist, and H. Moise, *J. Mol. Biol.*, **105**, 15 (1967).

8. M. O. Dayhoff, in *Chemical Evolution and the Origin of Life*, R. Buvet and C. Ponnamperuma eds., Elsevier, North Holland, Amsterdam, 1971, pp. 392.

9. D. O. Hall, R. Cammack, and K. K. Rao, in *Cosmochemical Evolution and the Origin of Life*, J. Oró, S. L. Miller, C. Ponnamperuma, and R. S. Young, eds., Reidel, Dordrecht, 1974, Vol. 1, pp. 363.

10. H. Baltscheffsky, in *Living Systems as Energy Converters*, R. Buvet, M. J. Allen, and J.-P. Massué, eds., Elsevier-North Holland, Amsterdam, 1977, pp. 81–87.

11. L. E. Orgel, *Isr. J. Chem.*, **10**, 287 (1972).

12. C. W. Carter, Jr., and J. Kraut, *Proc. Natl. Acad. Sci. U.S.A.*, **71**, 283 (1974).

13. G. Von Heyne, C. Blomberg, and H. Baltscheffsky, *Orig. Life*, **9**, 27 (1978).

14. H. Baltscheffsky, in *Energy Conservation in Biological Membranes*, G. Schäfer and M. Klingenberg, eds., Springer-Verlag, Berlin, 1978, pp. 3–18.

15. M. Baltscheffsky, *Arch. Biochem. Biophys.*, **133**, 46 (1969).

16. D. B. Knaff and J. W. Carr, *Arch. Biochem. Biophys.*, **193**, 379 (1979).

17. L. J. Prochaska and R. A. Dilley, *Biochem. Biophys. Res. Commun.*, **83**, 664 (1978).

18. M. Symons, C. Swysen, and C. Sybesma, *Biochim. Biophys. Acta*, **462**, 706 (1977).

19. M. Symons and A. R. Crofts, *Z. Naturforsch.*, **35c**, 139 (1980).

20. P. Mitchell, *Biol. Rev. Cambridge Phil. Soc.*, **41**, 445 (1966).

21. C. R. Woese and G. E. Fox, *Proc. Natl. Acad. Sci. USA*, **74**, 5088 (1977).

22. R. E. Fenna and B. W. Matthews, in *Chlorophyll-Proteins, Reaction Centers and Photosynthetic Membranes*, J. M. Olson and G. Hind, eds., Brookhaven Symposia in Biology Vol. 28, pp. 170–182 (1976).

THE ORIGIN AND
EVOLUTION OF LIFE AT
THE MOLECULAR LEVEL

MANFRED EIGEN

*Max-Planck-Institut für Biophysikalische Chemie,
Göttingen, Federal Republic of Germany*

I. BIOLOGICAL COMPLEXITY

The most conspicuous attribute of biological organization is its complexity. We see this especially clearly when we come down to the level of molecular detail. The physical problem of the origin of life can be reduced to the question: "Is there a mechanism by means of which complexity can be generated in a regular, reproducible way?" One way of regarding the origin of biological complexity as a process obeying natural law was put forward, 120 years ago, by Charles Darwin. His theses can be formulated in modern terms as follows:

Complex systems arise by an evolutionary process.

Evolution is based on natural selection.

Natural selection is a physical consequence of self-reproduction.

The third thesis is in fact neo-Darwinian. It follows from the quantitative treatments of population genetics developed, particularly by Haldane, Fischer, and Wright, in the first half of this century. The molecular-biological revolution of the 1950s led to the euphoric expectation that the laws of genetics would prove reducible to the magic formula

$$DNA \rightarrow RNA \rightarrow protein \rightarrow everything\ else$$

This dogma of molecular biology postulates that every detail of a complex structure is governed by information that flows irreversibly from the genotypic legislative to the phenotypic executive at the somatic level of an organism. In the 1980s, after the discovery of reverse transcriptases, restriction endonucleases, extrons, and introns, in short, of nature's tools for the processing of genotypic information, we phrase our statements a little more cautiously:

All organisms have to reproduce their genetic information.

Only nucleic acids can be reproduced true to sequence.

Reproduction provides a basis not only for the conservation of information but also for a selective processing of information and for optimization.

Thirty years of molecular-biological research have shown us how to ask the "right" questions. First of all, we must be able to observe accurately the basic process, viz., the reproduction of genetic information. From our results we construct an abstract, general biological principle with experimentally verifiable logical consequences. Finally, we search known biological structures for a "fossil" record, in order to find out whether the historical origin of life could in principle have followed the basic schema which we have formulated. Answers must be sought to the questions:

Can a causal connection be shown among molecular self-reproduction, selection, and evolution?

Is self-organization, based on self-reproduction and selection, an inevitable process, whose prerequisites and consequences can be found in natural systems?

Are there "fossil" relics of molecular evolution?

II. EXPERIMENTS IN MOLECULAR EVOLUTION

We shall first examine in detail the mechanism of reproduction of a virus whose genetic information lies encoded in a single-stranded RNA molecule. This virus is able to perform basically four functions, each of which is associated with a different protein unit: a capsid, which encloses the RNA and protects it from degradation by hydrolysis; a penetration enzyme, to inject the genetic material into the host cell; a factor to break down the host cell; a mechanism that reprograms the complex machinery of the host cell in accordance with instructions from the virus. In the case that we shall discuss, the last function is carried out by a protein molecule that associates with three ribosomal proteins to give an enzyme that recognizes the viral genome and replicates it very rapidly. The entire apparatus of metabolism and gene translation is turned over to the exclusive purpose of producing new virus particles. This factor thus embodies the essence of viral infection.

The reproduction mechanism of $Q\beta$, an RNA virus capable of infecting bacteria, is shown schematically in Fig. 1. During the replication of RNA,

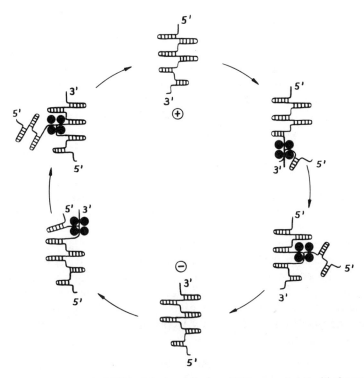

Fig. 1. The single-stranded RNA of the bacteriophage Qβ is reproduced with the assistance of an enzyme, called Qβ replicase, which consists of four subunits (black dots). The enzyme recognizes the matrix specifically and during synthesis, it moves from the 3' to the 5' end of the template strand. The replica formed (−) is complementary to the template (+). The 3' and 5' ends are symmetrically related in such a way that both plus- and minus-strands have similar 3' ends; both are recognized by the replicase, and the minus-strand thus acts as a template for the formation of a plus-strand. Internal folding of the two strands prevents the formation of a plus-minus double helix.

the enzyme, which consists of four subunits, moves from the 3' to the 5' end of the template. The newly formed RNA replica has an internally folded structure and this prevents the formation of a double strand. Sol Spiegelman,[1] who was the first to isolate the reproduction enzyme of this virus and, using it, to synthesize infectious viral RNA, has shown that annealing destroys the reproductive capacity of the virus. The reproduction does not take place at a steady speed: the enzyme pauses at so-called "pause sites." Presumably, it has to wait for the next portion of the matrix to melt apart before it can proceed to copy it. The discovery and isolation of a noninfectious RNA component, 220 nucleotides long, are also due to Spiegelman[2]; this RNA molecule, known as "midivariant," contains the 3' (and the 5') end needed for recognition by Qβ replicase and is therefore replicated by the enzyme, in fact much faster than the real Qβ RNA. "Midivariant" is thus a "scrounger," which cannot infect a cell alone, since it cannot produce the specific reproduction enzyme. In our laboratory the mechanism of RNA replication by Qβ replicase has been studied quantitatively.[3-6] The experiments were begun by Manfred Sumper and Rüdiger Luce and continued by Christof Biebricher and Rüdiger Luce. Our knowledge of the mechanism of the replication reaction results from a combination of (1) experimental investigation of the rate of replication as a function of the concentrations of substrate, enzyme, and RNA templates; (2) the mathematical analysis of a model for the replication (see Fig. 2); and (3) the computer simulation of this model using realistic values for parameters, obtained from experimental data.

A typical experiment is sketched out in Fig. 3. A ^{32}P-labeled nucleotide is used to measure the amount of newly synthesized RNA as a function of time. The concentrations of the substrate (the four nucleoside triphosphates of A, U, G, and C) and that of the enzyme are constant. The initial template concentration is varied by serial dilution with a constant dilution factor. The increase in RNA concentration in the course of time can be divided into three phases:

1. An induction period, which increases logarithmically with increasing dilution. Repeated dilution by a constant factor produces a series of constant displacements along the time axis.
2. A linear increase in RNA concentration, starting at the moment when the RNA concentration becomes (approximately) equal to the enzyme concentration.
3. A plateau, which is not reached until the concentration of the RNA greatly exceeds that of the enzyme.

This behavior can be explained by the following kinetic considerations. The reaction leading to new RNA templates is catalyzed by a complex of enzyme and template. The affinity between these partners is so high that at the concentrations used every RNA molecule binds to an enzyme. The number of catalytically active complexes then rises exponentially until the RNA concentration becomes equal to the concentration of enzyme. At this point the enzyme is saturated with RNA. From now on the number of catalytically active RNA–enzyme complexes remains constant and the synthesis enters the linear phase, that is, the rate of appearance of new RNA molecules becomes constant. The new RNA molecules, now in excess over the enzyme, bind not only as templates but also, less strongly, at the site of synthesis. This leads to inhibition of synthesis by

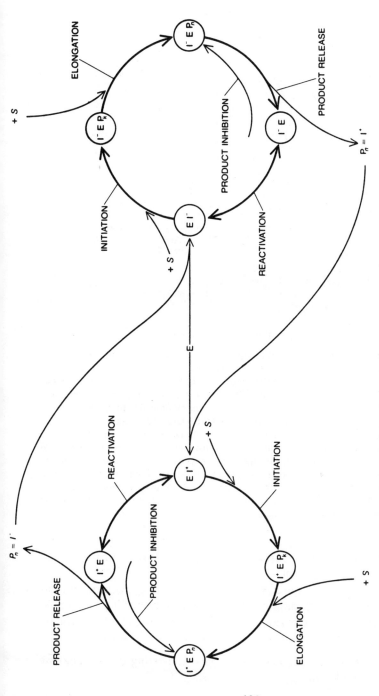

Fig. 2. A characteristic of the mechanism of RNA replication is the coupled pair of cycles of synthesis, for the plus- and minus-strand, respectively. A catalytically active complex consists of the enzyme (replicase) and an RNA template. Four phases of each cycle can be distinguished: (1) the commencement of replication by the binding of at least two substrate (nucleoside triphosphate) molecules; (2) elongation of the replica strand by successive incorporation of nucleotides; (3) dissociation of the complete replica away from the replicase; (4) dissociation of the enzyme from the 5' end of the template and its reassociation with the 3' end of a new template. The matrix is represented by I (information), the enzyme by E, and the reaction product by P. The ultimate reaction product P_n is then used as a template (I). The substrate, S, is the triphosphate of one of the four nucleosides A, U, G, and C.

123

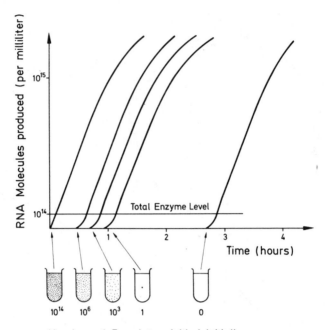

Number of Templates Added Initially

Fig. 3. If a synthesis mixture (buffer, salts, Qβ replicase, nucleoside triphosphates, and RNA templates) containing equal numbers of enzyme molecules and template strands is incubated, then the rate of production of RNA molecules is linear in time, since the number of catalytically effective complexes is equal to the (constant) number of enzyme molecules. (The flattening of the reaction curves at high RNA concentration is due to inhibition of the enzyme by binding of excess RNA to the active site.) If the template concentration is lowered in steps of a particular factor, the growth curves are displaced along the time axis by equal intervals. This logarithmic dependence of the induction period indicates an exponential growth law, which holds as long as the enzyme is in excess over the template RNA, during which time the number of catalytically active complexes perpetually increases (autocatalysis). Even if no template is present at the beginning, an RNA arises *de novo* after a long induction period. This acts as a template and multiplies rapidly. Such synthesis of RNA *de novo* is a particular property of Qβ replicase.

the excess RNA molecules present in the linear phase, and finally the synthesis comes to a standstill.

By altering the reaction conditions (e.g., substrate concentration, enzyme concentration, initial ratio of plus- to minus-strand) it proved possible to measure the kinetic parameters and to verify the reaction scheme of Fig. 2. The principal conclusions of this investigation were

The replication is catalyzed by a complex consisting of one enzyme and one RNA molecule.

The growth rate is proportional to the lower of the two total concentrations (enzyme or RNA). This means for low and high RNA concentrations exponential and linear growth, respectively.

If two different mutants compete, then even in the linear region the mutant with the selective advantage continues to grow exponentially, until this mutant saturates the enzyme.

A selective advantage in the linear phase is given solely by differences in the kinetics of enzyme–template binding. In the exponential phase, in contrast, the competition is based on overall rates of production.

The overall growth rate of plus- and minus-strands is given in the exponential phase by the geometric mean of their rate parameters and in the linear phase by the harmonic mean. The ensemble of plus- and minus-strands grows up in the ratio of the square roots of their respective rate parameters of production.

The replication rate depends on the length of the RNA chain to be synthesized and on the concentrations of the substrates (A, U, G, C). The latter dependence is weaker than linear as successive substrate molecules are frequently prebound to the enzyme.

We now arrive at the actual question which we wished to answer: What are the consequences of the laws of replication? Can new properties be attained by the evolution of a replicating system? Are the characteristics of the replicative system sufficient, or do we need to look for additional necessary properties?

Spiegelman[7] has provided an important stimulus for research in this direction. By serial transfer of RNA templates from one nutrient glass to another he obtained, at the end of such a series, selected variants of the Qβ genome which were no longer infectious but which showed a higher replication rate (measured per nucleotide). They made use of their higher replication rate to escape the selection pressure of dilution. The significant finding, however, was that these new variants not only had a higher rate of chain elongation but also possessed 500 instead of the original 4500 nucleotides, which enabled them to complete a round of replication in a fraction of the original time. Evolution of this sort, resulting in a loss of information, might well rather be termed "degeneration." However, the experiments showed that this replication system is highly capable of adaptation, an indispensable prerequisite for evolutionary behavior.

These experiments became particularly relevant when, in 1974, a surprising observation was made by Manfred Sumper. In series which were diluted so far that in each sample the probability of finding a template molecule was very small, there arose, nonetheless, reproducibly and homogeneously, a molecule with size and structure similar to the "midivariant" isolated by Spiegelman. This phenomenon is indicated in Fig. 3. In contrast to the template-instructed replication, this synthesis was associated with a disproportionately long induction period, which depended critically on the reaction conditions. Sumper was convinced that he had found a variant which the Qβ enzyme had "invented" and synthesized *de novo*. (Other experts in the field believed almost unanimously that he was seeing an impurity carried by the enzyme.) Sumper was able by further experiments to refute the "impurity" postulate, and since then the existence of synthesis *de novo* has been amply proved, above all in experiments conducted by Christof Biebricher and Rüdiger Luce. Kinetic measurements have shown that the induction of template-instructed and *de novo* synthesis are subject to entirely different rate laws. Thus, for example, the template-instructed synthesis proceeds on a single enzyme molecule and the substrates are introduced successively into the growing chain. Synthesis *de novo*, on the other hand, requires a complex of several enzyme molecules, and the rate-determining step is a nucleation by three or even four substrate molecules. However the principal evidence lies in the demonstration that, in the early phase of synthesis, variants appear which lengthen under selection pressure and which lead under different experiment conditions to different end products. The last observation is also due to Manfred Sumper. He obtained different "minivariants," which grew normally under conditions in which the wild-type was no longer capable of existence, for example, in the presence of reaction inhibitors.

The decisive experiment of Biebricher and Luce is shown in Fig. 4. A synthesis medium containing highly purified enzyme and substrates is incubated and maintained at a suitable temperature, for a time adequate to allow the multiplication of any templates present but too short to enable products to arise *de novo*. Then the solution is divided into portions. Each portion is incubated long enough to allow synthesis *de novo* and the products are compared by the fingerprint method. If the "impurity" hypothesis is correct, then multiplication of the impurity in the first phase should lead to the same product from each portion of the incubated medium. If the *de novo* hypothesis is correct then the products should be different, since at the beginning different enzyme molecules were working on different products. Selection, that is, preferential reproduction of *one* rudimentary strand, could not yet take place, since in the first, short incubation none of the products *de novo* was complete.

The experiment gave many different products. Only if these were mixed and again incubated did a single, homogeneous, reproducible variant grow up. The earliest products that could be detected directly were about 70 nucleotides long. In the course of evolution there appeared longer chains,

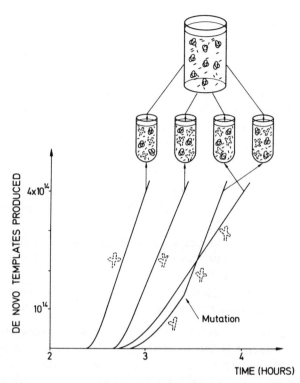

Fig. 4. A solution of nucleotide triphosphate is incubated in the presence of Qβ replicase for just long enough to assure the manifold replication of any templates that may contaminate the enzyme. The incubation is interrupted before even one template has time to arise *de novo*. The solution is then divided up into portions and the incubation is continued, this time long enough to allow products to arise *de novo* and to multiply. The RNA formed in each portion is analyzed by the fingerprint method; various different reaction products are found. Sometimes the growth curve displays the appearance of a new mutant. Although the incubation time of template-instructed synthesis is determined unambiguously (because of the superposition of many individual processes) the synthesis *de novo* shows a scatter of induction times. This indicates that the initiation step is a unique molecular process which is then rapidly "amplified."

for example, at high salt concentration the "midivariants," with about 220 nucleotides.

The important result of these experiments is, however, not just an explanation of this unusual feature of the Qβ system. More important, we now have a flexible replication system at our disposal, with which a series of further interesting studies can be set up. Above all, it can be shown that selection and evolution are inevitable consequences of self-replication, and can as such be investigated quantitatively. For example, the question of rapid optimization under extreme experimental conditions has been answered in detail. The results of the evolution experiments described can be summarized in four principal statements:

The synthesis *de novo* of RNA by Qβ replicase proceeds by a mechanism fundamentally different from that of template-instructed RNA synthesis. The active reaction complex contains at least two enzyme molecules and requires nucleation by a seed of three or four substrate molecules.

The rate-limiting step is nucleation, and elongation and reproduction follow rapidly. The singular nature of the molecular process which initiates the reaction is reflected in the scatter in induction times for synthesis *de novo*.

Synthesis *de novo* produces a broad spectrum of mutants of varying length, containing sequences capable of adaptation to a great variety of environmental conditions.

Initiation of self-reproduction is clearly sufficient to set the process of evolutive optimization in motion.

This finding allows the development of an evolution reactor, in which optimally reproducing RNA sequences may be produced in a relatively short time. This in turn makes possible the development of a principle for the evolution of RNA structures with optimized translation products. Experiments in this direction are in progress.

Self-replication and mutagenicity in an open system far from equilibrium are thus sufficient to produce behavior patterns including selection and evolution. Even in relatively simple replication systems properties optimal with respect to the wild-type can be produced *in vitro* in a few generations. Such effects must be the consequence of a physical principle. Can such a principle be formulated quantitatively?

III. SELECTION AND EVOLUTION GOVERNED
BY NATURAL LAW

In earlier publications we have shown that the principle of selection can be deduced from the premises of a self-replicating system as an extremum principle. It states that inherent linear autocatalysis causes the relative population numbers to take on values that correspond to the highest reproductive efficiency of the system as a whole. The distribution of relative concentrations in the stationary population is, after a short induction period, independent of changes of the system as a whole. The population consists of a uniquely defined wild-type (or several equivalent, i.e., "degenerate," variants) and a spectrum of mutants. The wild-type is most frequently represented in the distribution, but in a well-adapted population it comprises only a small fraction of the total. The quotient of the population number (x_i) of the individual mutant i and that of the wild-type (x_m) is given by a function of the rate parameters for mutation (W_{im}, for $m \rightarrow i$) and reproduction (W_{mm} and W_{ii} for $m \rightarrow m$ and $i \rightarrow i$, respectively):

$$\frac{x_i}{x_m} = \frac{W_{im}}{W_{mm} - W_{ii}} \tag{1}$$

Supposing mutant i arises solely from wild-type m, the following variables are also of importance:

$\bar{q}$ = average accuracy of copying of a nucleotide
$1 - \bar{q}$ = average error rate per nucleotide
v_i = number of nucleotides in the sequence i
$Q_i \approx \bar{q}^{v_i}$ = fraction of sequences of type i correctly copied
σ_m = superiority of the wild-type over its spectrum of mutants (corresponding in general to the ratio of the replication rate of the wild-type and the average replication rate of the mutants)

The consequences of this extremum principle, valid for Darwinian systems, are

Selection of a distribution of mutants dominated by the wild-type. This is only stable as long as the conditions $\sigma_m > 1$ and $Q_m > \sigma_m^{-1}$ are fulfilled.

Evolution by the selection of newly appearing mutants, which by virtue of a selective advantage disobey the condition $\sigma_m > 1$ and thus destabilize the dominance of the wild-type.

Restriction of the information content due to the conditions $Q_m > \sigma_m^{-1}$. The upper limit is given by $v_{max} = \ln (\sigma_m)/(1 - \bar{q}_m)$. It corresponds roughly to the reciprocal of the average error rate ($1 - \bar{q}_m$), as long as σ_m is sufficiently larger than unity. If the information content v_m of the wild-type approaches the upper limit v_{max}, Q_m becomes approximately equal to σ_m^{-1}. The proportion of wild-type in the total population is then very small:

$$\frac{x_m}{\sum\limits_{k} x_k} = \frac{Q_m - \sigma_m^{-1}}{1 - \sigma_m^{-1}} \tag{2}$$

Both selection (the stabilization of a particular distribution) and evolution (the establishment of one new population after another) result from an "inward compulsion." They are the inevitable consequences of self-reproduction behavior.

The fact that the evolutionary optimization process indeed reaches the "mountain peaks" and does not get stuck in the "foothills" lies in the topology of multidimensional mutation space. Consider for example a binary sequence with v members. We can give each position in the sequence a coordinate axis, with two points and thus obtain a v-dimensional phase space in which each of the 2^v points represents a mutant. The process of evolution can then be regarded as a route in this space, characterized by a continually rising selection value. The topology of such a multidimensional space is not easily imaginable; the "mountains" are extremely bizarre, for although these are 2^v points the greatest separation (in terms of mutation steps) is only v. There are saddle points of various orders, at which movement in k directions leads uphill and in $(v - k)$ directions downhill. This provides a sufficient basis for relatively small mutational jumps to enable the system always to find an uphill route. There is a best value of v, for which the number of routes is large enough and the probability of multiple mutations (depending on population size) is great enough for any optimal "peak" to be reached.

Let us summarize: selection, evolution, and optimization are processes that follow regular physical laws and that can be formulated quantitatively. This of course does not mean that the actual, historical process of evolution can be deduced from theory. The starting point, the complex boundary conditions, and the multitude of superimposed perturbing influences are all more or less unknown. Theory tells us simply what follows when certain premises are set up and certain boundary conditions are imposed. It explains the reproducible, regular phenomenon of nature in an "if–then" description. This generalization applies to the theory outlined here, which has helped us to interpret and explain the experimental results described above. It has been confirmed by quantitative measurement under the exactly defined initial and boundary conditions of the laboratory. For processes occurring in nature, however, it reveals only trends, minimal requirements, limitations and, perhaps, some consequences. It must also be shown by experimental investigation whether conclusions from the theory have any relevance for naturally occurring processes. Two such conclusions are especially worthy of mention:

The quantity of information which can be selected in a molecular population depends on the average error rate and the average selective advantage of the wild-type. Crossing the critical error threshold leads to such an accumulation of errors that the information in the wild-type sequence is irretrievably lost.

The capacity of the wild-type for adaptation is greatest close to the error threshold. The quantity of information compatible with a stable distribution is then in optimal relation to the variety in its spectrum of mutants. Such a system responds very flexibly to changes in its environment. The wild-type is the predominant individual sequence, but it makes up only a small fraction of the complete mutant spectrum.

IV. MOLECULAR RECORDS OF EVOLUTION IN NATURE

The predictions of the theory of evolution can be tested on natural systems. Charles Weissmann and his co-workers[10,11] have obtained the following results for Qβ viruses. The wild-type has a defined sequence, which, however, does not mean that the majority of viruses share exactly the same sequence. It means merely that the superposition of all sequences gives an unambiguous "majority" or master sequence, namely, that of the wild-type.

The cloning of single viruses or single viral RNA molecules followed by their rapid multiplication leads to populations with various sequences. The sequences generally deviate in one or two positions from the wild-type, which is itself found in hardly any clone (see Fig. 5). The fact that the wild-type makes up only a (nearly negligibly) small fraction of the mutant spectrum implies that the information content of the wild-type has very closely approached the threshold value v_{max}.

Deliberately produced extracistronic single mutations (these are nonlethal mutations in portions of the sequence that are not translated) revert to the wild-type. They correct their errors with a probability around 3×10^{-4}. Quantitative analysis of such data allows the determination of the error rate, the selective advantage of the wild-type, and thus the establishment of the critical error threshold for the transmission of the information in the genome. This value agrees within experimental error with the quantity of information present (4500 nucleotides). The fact that the cloning of individual mutants is possible at all is due to the dominance in the distribution, around the wild-type, of mutants whose growth rates are very similar to that of the wild-type itself. These are preferentially "fished out" in the serial dilution steps needed for the cloning of single molecules. Since they multiply nearly as rapidly as the wild-type, they begin by producing a spectrum of mutants with an average sequence identical to their own. At some point, the wild-type will reappear in this spectrum. However, it can only assert itself slowly, the speed with which it does so corresponding to the (small) difference between the growth rates of

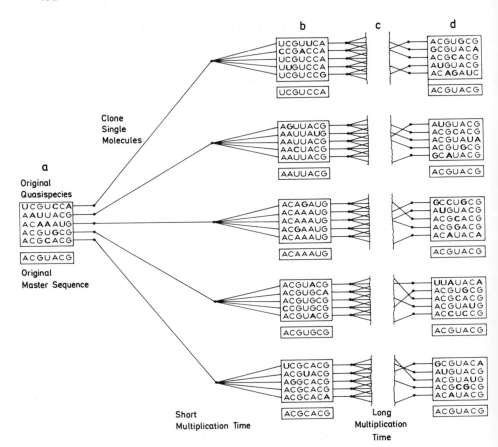

Fig. 5. In the experiment, carried out by Charles Weissmann and his co-workers, single QβRNA molecules (or viruses) from a wild-type distribution (*a*) were cloned in *E. coli* bacteria. After rapid multiplication the clones of individual RNA molecules (*b*) were analyzed and compared by the fingerprint method. Differences were noticed in one or two positions in the sequence. After a further, long period of reproduction (*c*) the wild-type distribution (*d*) was found in every clone, that is, the average sequences had become identical again.

the wild-type and the cloned mutant. Finally each clone is dominated by the wild-type (see Fig. 5).

Yet again the conclusion can be drawn that all single-stranded RNA viruses are subject to similar restrictions with regard to information content. In nature there are no (single-stranded) RNA viruses whose replicative unit contains more than the order of 10^4 nucleotides. All larger viruses possess double-stranded nucleic acids or are composed of several replicative units. These in turn are subject to analogous relationships between the error threshold and the maximum reproducible information content. DNA polymerases in general work more accurately than RNA polymerases. This is due to their additional facility for recognition and correction of errors.

What do these results signify for our understanding of early evolution?

The first replicative units must have possessed considerably less information than the RNA viruses, which work with an optimized RNA-copying machinery. In the absence of efficiently adapted enzymes the accuracy of reproduction depends solely on the stability of the base pairs. Under these conditions the GC pair has a selective advantage over the AU pair of a factor of about 10. Model experiments show that for GC-rich polynucleotides the error rate per nucleotide can hardly be reduced below a value of 10^{-2}. The first "genes" must accordingly have been polynucleotides with a chain length around 100 bases or less.

Molecular evolution demands inherent self-reproductivity. RNA seems to fulfill this function best of all known macromolecules. On account of its complex structure RNA must first have appeared in nature long after proteins or protein-like structures. A protein can by chance fulfill a particular function, but this fulfilment is determined by purely structural and not at all by functional criteria. Adaptation to a particular function, however, demands an inherent mechanism of self-reproduction. The only logically justifiable way of exploiting the immense functional capacity of the proteins in evolution lies in an intermarriage between these two classes of macromolecules, that is, in the translation into protein of the information stored in the self-reproductive RNA structures.

This at once raises the question: "Could RNA ever have arisen without the help of enzymes, without replicases?" Experiments by Leslie Orgel and his co-workers[12] suggest that this was possible. It was found that zinc ions, found today as cofactors in all replicases, are excellent catalysts for the 3'-5' union of nucleotides, thus allowing the template-instructed synthesis of polymers. This was first demonstrated with poly-C as template. If activated G and A nucleotides are offered, in equal concentration, then G is preferentially incorporated into the product by a factor, depending on reaction conditions, between 30 and 200.

This suggests strongly that in a suitable medium GC-rich strands with a chain length around 100 nucleotides will arise spontaneously, reproduce themselves, and undergo adaptation by evolution: Can we today find a record of these first "genes?"

The information content of such "genes" suffices only for relatively small proteins, certainly not optimally adapted. This means, however, that these are by now long outdated as information carriers and have been displaced by better ones. The displacement proceeded hand in hand with the development of the machinery of translation. In the translation apparatus RNA structures carry not only information but also their own function and in addition they represent targets for processing functions. It is therefore more likely that they have survived up to the present as functional entities, for example as transfer RNA (tRNA) or as ribosomal RNA (rRNA) than as message carriers. Since the functional RNA molecules did not have to store genetic information, they were hardly subject to selection pressure once they had arrived at an adapted structure. The functional nucleic acids were recruited to begin with from the same reservoir as their information-carrying sister molecules, the mRNAs. We may thus expect for example tRNA, as an eyewitness of the early evolution of the translation apparatus, to retain some memory of the structure of the earliest "genes." This molecule, with a chain length of about 76 nucleotides, fulfills exactly the criteria that theory requires, which present-day structures confirm and which are relevant for prebiotic conditions.

Many tRNA sequences are known today, both for a given codon at a variety of phylogenic levels and for a given species and many different anticodons. Each such category is interesting[13,14] for comparative analysis. Phylogenic analysis shows whether tRNA has retained information from prebiotic times, or this information has been lost in the course of evolution. The comparison of different tRNA molecules in a single species may then lead to a reasonably complete reconstruction of the early forms and allow statements about the early evolution of the translation apparatus.

The phylogenic family tree of $tRNA_{init}^{Met}$ in Fig. 6 shows that tRNA belongs to the most highly conserved structures that we know. Of the examples shown the tRNAs of the fruit fly (*Drosophila*) and the starfish differ only in a single nucleotide pair, and these differ from human tRNA at only four (weighted) positions. Organisms that parted company thousands of millions of years ago, such as the eubacteria, the chloroplasts, and the archaebacteria, appear on the tRNA scale as close relations, with closely connected sequences.

This brings a reconstruction of the primeval structure into the realm of the possible. If the sequences of different tRNAs, for example, for *E. coli* bacteria, yeast cells, or archaebacteria, are compared, then they all show a high GC content, which increases further when the sequences are superimposed. However, it can be seen from the mitochondrial sequences, which are all rich in A and U, that G and C could be displaced in the course of evolution, presumably on account of a rich supply of the metabolite A in mitochondria; G and C are thus not demanded by con-

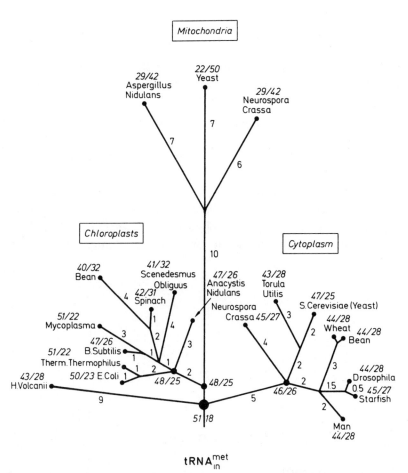

$$\text{tRNA}^{\text{met}}_{\text{in}}$$

Fig. 6. The family tree of a tRNA, here the one involved in the initiation of translation, shows a few changes in nucleotide sequence (indicated by numbers of the branches), even after thousands of million of years. All such known mammalian sequences are practically identical. The quoitents give the ratio of (guanidine plus cytosine) to (adenine plus uracil). This ratio is greatest near the earliest branching points and smallest at the ends of the long branches; it is almost reversed in the mitochondria (the "power stations" of the cell), which have a ratio around 1:2, vis-à-vis the early branching points, with a ratio around 2:1. Four groups emerge clearly: the archaebacteria (only one representative, *H. volcanii*), the eu-bacteria and blue algae (hardly distinguishable from the chloroplasts), the eukaryotes, and the mitochondria. The long distance to the mitochondria reflects a high rate of replacement of G and C by A and U. The purine-pyrimidine succession shows the mitochondria to be close relatives of the eubacteria, whereas their distance from the archaebacteria and the eukaryotes is relatively large.

135

siderations such as stability. Further, the reconstructed precursor sequences suggest a periodic triplet structure (see Fig. 7), which in turn suggests a primeval code pattern GNC (N = any of the four nucleotides A, U, G, and C). Thus, tRNA was not only the primitive adaptor but also a primitive carrier of genetic information, a function now lost in the course of evolution. A comparative analysis of tRNA sequences leads to conclusions that can be summarized as follows:

tRNA is an "ancient" adaptor, which has changed relatively little in the course of phylogenesis.

Different tRNA molecules within a species appear as mutants of a master sequence.

The original master sequence is largely capable of reconstruction.

It is characterized by a high GC content and a code pattern RNY.

All findings to date are compatible with a primeval code GNC for the amino acids most common in nature: glycine, alanine, aspartic acid, and valine.

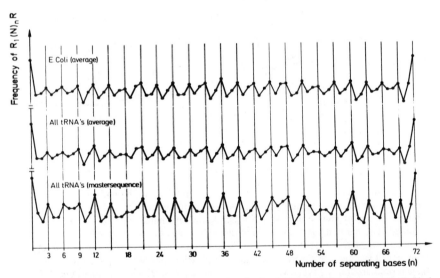

Fig. 7.   Correlation analysis of the repetition of purine in tRNA. A tRNA sequence is divided into triplets, beginning at the 5' end and in phase with the anticodon. The frequency with which a purine (R) in the first position of the triplet occurs n positions later is counted and plotted against n. The period of three which emerges indicates clearly a triplet structure of the form RNY. The curves show values for the averaged sequences of *E. coli* and of all tRNAs investigated to date; these are compared with that of the master sequence arising from the superposition of all tRNAs. The fact that the correlation is clearer in the master sequence suggests that this may represent a "memory" of the earliest phase of evolution.

V. SYNOPSIS

We indeed find a congruence between theory, model experiment, and "historical" record. Irrespective of this, we should still pay attention to the following caveats: The physical theory of evolution, like every other physical theory, describes no more than an "if–then" behavior pattern. If the theory is correct then it predicts the consequences resulting from particular initial conditions.

The value of the theory is to be assessed exclusively by its capacity for experimental test. Model experiments give quantitative "standards" by means of which the probability of each stage in the emergence of life may be estimated.

Neither theory nor experiment allows a conclusion about the actual historical process of evolution. This requires a specific historical record.

The congruence between theory, model experiment and historical record enables us to regard the principle of "life" as one of nature's regularities.

On the basis of these principles the evolution process may be simulated and reproduced in the laboratory.

Acknowledgments

I am most grateful to Dr. Ruthild Winkler-Oswatitsch for her help in preparing the manuscript and to Dr. Paul Woolley for translation of the manuscript into English.

References

1. S. Spiegelman et al., *Proc. Natl. Acad. Sci. USA,* **50,** 905 (1963); **54,** 579, 919 (1965); **60,** 866 (1968); **63,** 805 (1969).

2. D. R. Mills, F. R. Kramer, and S. Spiegelman, *Science,* **180,** 916 (1973); F. R. Kramer and D. R. Mills, Proc. Natl. Acad. Sci. USA, **75,** 5334 (1978).

3. M. Sumper and R. Luce, *Proc. Natl. Acad. Sci. USA,* **72,** 162 (1975).

4. C. K. Biebricher, M. Eigen, and R. Luce, *J. Mol. Biol.,* **148,** 369 (1981).

5. C. K. Biebricher, M. Eigen, and R. Luce, *J. Mol. Biol.,* **148,** 391 (1981).

6. C. K. Biebricher, M. Eigen, and W. C. Gardiner, Jr., *Biochemistry,* **22,** 2544 (1983).

7. D. R. Mills, R. I. Peterson, and S. Spiegelman, *Proc. Natl. Acad. Sci. USA,* **581,** 217 (1967).

8. M. Eigen, *Naturwissenschaften,* **58,** 465 (1971).

9. M. Eigen and P. Schuster, *Naturwissenschaften,* **64,** 541 (1977); **65,** 7, 341 (1978).

10. E. Domingo, R. A. Flavell, and C. Weissmann, *Gene,* **1,** 3 (1976); E. Batschelet, E. Domingo, and C. Weissmann, *Gene,* **1,** 27 (1976).

11. C. Weissmann, G. Feix, and H. Slor, *Cold Spring Harbor Symp. Quant. Biol.,* **33,** 83 (1968).

12. R. Lohrmann, P. K. Bridson, and L. E. Orgel, *Science,* **208,** 1464 (1980); *J. Mol. Evol.,* **17,** 303 (1981); P. K. Bridson and L. E. Orgel, *J. Mol. Biol.,* **144,** 567 (1980).

13. M. Eigen and R. Winkler-Oswatitsch, *Naturwissenschaften,* **68,** 217 (1981).

14. M. Eigen and R. Winkler-Oswatitsch, *Naturwissenschaften,* **68,** 282 (1981).

COMMENTARY: ON PAPER BY MANFRED EIGEN

ROBERT M. BOCK

The Graduate School, University of Wisconsin, Madison, Wisconsin

The quantitative relations which Dr. Eigen has presented are an elegant and appropriate framework on which to assemble the specific molecular theories of evolution of a self-reproducing system. For many years I have taught my students of the likelihood that the primitive self-reproducing systems were of low information content, low specificity, and tolerant of limited classes of errors. Polymers of uracyl and inosine permit a two letter (purine-pyrimidine) code which is tolerant of incorporation of small numbers of cytidine, adenosine, and guanosine without loss of information. The gradual introduction of specific GC and AU pairs adds information content and permits increased specificity. The U-I polymers are capable of three-base and four-base interactions, which could have provided early molecular clusters with affinity for amino acids or ribonucleotides. The driving forces for polymerization were cyclic dehydration on daily or even annual cycles. The many attractive features of the primitive purine-pyrimidine code have led me to propose that "in the beginning there were U and I."

I propose that the addition of proteins to the evolving self-reproducing system is an early step analagous to our change from a hunter-gatherer society to a tool-using society. The tool of polymeric amino acids was so powerful that it replaced all less efficient catalysts and any substrate-gathering nucleotide clusters. Thus, transfer RNA was not a part of the most primitive gene but probably appeared as soon as amino acid tools became important in catalysis or in stabilization of polynucleotides.

Professor Eigen's parameters of excess production, specificity, and error frequency are very useful in describing the process of change from a U-I code to the four letter triplet code we know today.

The evolutionary change, from a purine-pyrimidine code with only the first two letters encoding sense, to a triplet code is a further evolutionary development which fits Eigen's predicted progress toward increasing specificity/efficiency and toward decreasing tolerance of errors. In the

primitive purine-pyrimidine code each third base satisfied a spacing requirement but contributed negligible specific information. The triplet code
we observe in nature today derives substantial information from the third
nucleotide in the codon although that position is clearly less specific than
the first two bases of the triplet. Thus, there are several simultaneous
processes in the gradual evolution of the nucleic acid code to reach the
specificity and complexity evident in biology today.

OPTIMIZATION OF
MITOCHONDRIAL ENERGY CONVERSIONS

J. W. STUCKI

Pharmakologisches Institut der Universität Bern, Bern, Switzerland

I. INTRODUCTION

In aerobic tissue oxidative phosphorylation is the most important source of the energy rich phosphate ATP. In this process the free energy resulting from the combustion of oxidizable substrates is converted into the phosphate potential, which is then used to drive the energy-requiring processes occurring in the living cell. Since the process of oxidative phosphorylation is localized within the mitochondria, these organelles have often been apostrophed to be the powerhouse of the cell. To be more specific, the mechanism of oxidative phosphorylation is intimately integrated within the inner mitochondrial membrane. For this reason oxidative phosphorylation can be regarded as a membrane-linked energy converter. This paper outlines several principles by which the performance of this energy converter can be optimized.

A quantitative description of oxidative phosphorylation within the cellular environment can be obtained on the basis of nonequilibrium thermodynamics. For this we consider the simple and purely phenomenological scheme depicted in Fig. 1. The input potential X_0 applied to the converter is the redox potential of the respiratory substrates produced in intermediary metabolism. The input flow J_0 conjugate to the input force X_0 is the net rate of oxygen consumption. The input potential is converted into the output potential X_p which is the phosphate potential $X_p = -[\Delta G_{phos}^{o\prime} + RT \ln(\text{ATP/ADP·}P_i)]$. The output flow J_p conjugate to the output force X_p is the net rate of ATP synthesis. The ATP produced by the converter is used to drive the ATP-utilizing reactions in the cell which are summarized by the load conductance L_l. Since the net flows of ATP are large in comparison to the total adenine nucleotide pool to be turned over in the cell, the flow J_p is essentially conservative.

The relations between flows and forces in oxidative phosphorylation

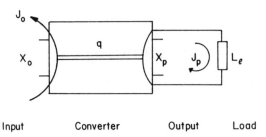

Input   Converter   Output   Load

Fig. 1. Input–output relationship of oxidative phosphorylation with attached load. For meanings of symbols see text.

can be described by the linear phenomenological laws

$$J_p = L_p X_p + L_{p0} X_0 \tag{1}$$

$$J_0 = L_{p0} X_p + L_0 X_0 \tag{2}$$

where the L's are phenomenological coefficients. In addition to linearity also reciprocal relations are assumed in this description, that is, $L_{p0} = L_{0p}$. With these assumptions equations (1) and (2) have the same mathematical structure as generally given by linear nonequilibrium thermodynamics.

The experiment depicted in Fig. 2 was designed to test the validity of the two fundamental assumptions introduced above, that is, linearity and reciprocity. Isolated mitochondria from rat liver were incubated in the presence of a constant input force X_0.[1] The output force X_p was varied within the physiologically relevant range through additions of glucose plus hexokinase. A linear regression analysis of the measured flows J_p and J_0 as a function of X_p showed that linearity as well as reciprocity were fulfilled within experimental error.

This experimental verification entitles us to apply the theory of linear energy converters to oxidative phosphorylation.[2] Kedem and Caplan have introduced a useful normalization of the straight and the cross coefficients of the scheme [equations (1) and (2)].

$$Z = \sqrt{L_p/L_0} \tag{3}$$

is the phenomenological stoichiometry and

$$q = L_{p0}/\sqrt{L_p L_0} \tag{4}$$

is the degree of coupling of oxidative phosphorylation. In the region where oxidation drives phosphorylation we have $0 < q < 1$.

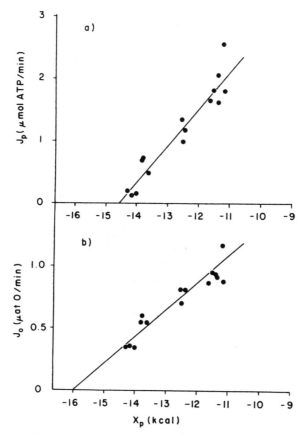

Fig. 2. Linear and symmetric relations between flows and forces in oxidative phosphorylation. 7.2 mg isolated mitochondria from rat liver were incubated in 3 ml of a medium consisting of 100 mM KCl, 5 mM potassium phosphate pH 7.4, 5 mM potassium glutamate pH 7.4, 5 mM potassium malate pH 7.4, and 1 mM ATP at 37°C. When the oxygen tension in the closed vessel had dropped to 75% of air saturation 6.67 mM glucose and varying concentrations of hexokinase (0.35, 0.7, 3.5, 7.0, and 35.0 units) were injected into the medium. The incubations were stopped by Millipore filtration when the oxygen tension had reached 15% of the original value. J_o was measured with a Clark oxygen electrode and J_p as well as X_p were calculated from the measured steady state concentrations of ATP, ADP, AMP and from the measured accumulated glucose 6-phosphate. X_o was estimated as 45 kcal/mole during the incubation. A linear regression analysis of the plots of flows versus X_p yielded the following phenomenological coefficients: $L_p = 0.616$, $L_{po} = 0.198$ with a regression coefficient of 0.95 and $L_{op} = 0.214$, $L_o = 0.0762$ with a regression coefficient of 0.94. For further details see ref. 1.

The quantity of main interest for our considerations is the efficiency function[2]

$$\eta = -\frac{J_p X_p}{J_0 X_0} = -\frac{(q + x)x}{qx + 1} \tag{5}$$

where $x = Z X_p / X_0$ is the force ratio. The efficiency plotted as a function of the force ratio reveals two steady states with zero efficiency (Fig. 3). In between these states efficiency passes through an optimum. This optimal efficiency

$$\eta_{opt} = \frac{q^2}{(1 + \sqrt{1 - q^2})^2} \tag{6}$$

as well as the force ratio permitting optimal efficiency

$$x_{opt} = -\frac{q}{1 + \sqrt{1 - q^2}} \tag{7}$$

are functions of the degree of coupling only.

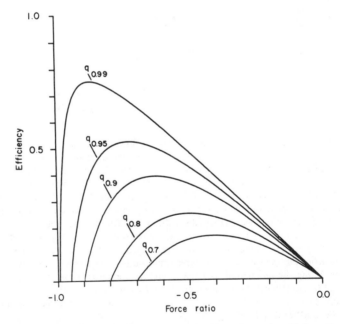

Fig. 3.   Dependence of efficiency on the force ratio. Plot of equation (5) for values of the degree of coupling indicated in the figure.

The zero efficiency state characterized by a vanishing output flow (J_p = 0) is called static head. At this steady state the emphasis of the machinery of oxidative phosphorylation is the maintenance of a maximal phosphate potential rather than a net rate of ATP synthesis. All the energy is expended in order to maintain that potential and to compensate the losses due to incomplete coupling, that is, $q < 1$. This situation is reminiscent of the operation of a refrigerator, which maintains a low temperature by compensating the leaks due to imperfect insulation. Physically the static head situation corresponds to the open circuited operation of the energy converter or, in other words, to the operation of the energy converter with an attached zero load conductance. The other extreme mode of operation of the energy converter results from attaching an infinite load conductance. This short circuited operation leads to the other zero efficiency state of the system namely the one characterized by a vanishing output force ($X_p = 0$). This state has been called level flow.

In order to extract the maximal energy out of the available foodstuff oxidative phosphorylation should operate at the state of optimal efficiency *in vivo*. Since a zero as well as an infinite load conductance both lead to a zero efficiency state, obviously there must be a finite value of the load conductance permitting the operation of the energy converter at optimal efficiency. For linear thermodynamic systems like the one given in equations (1) and (2) the theorem of minimal entropy production at steady state constitutes a general evolution criterion as well as a stability criterion.[3] Therefore, the value of the load conductance permitting optimal efficiency of oxidative phosphorylation can be calculated by minimizing the entropy production of the system (oxidative phosphorylation with an attached load)

$$\frac{d_iS}{dt} = \left[x^2\left(1 + \frac{L_l}{L_p}\right) + 2qx + 1 \right] L_0 X_0^2 \qquad (8)$$

at the force ratio x_{opt} [equation (7)] where efficiency is optimal. It is easy to see that d_iS/dt assumes a minimum at x_{opt} if and only if

$$\frac{L_l}{L_p} = \sqrt{1 - q^2} \qquad (9)$$

This necessary and sufficient condition called *conductance matching of oxidative phosphorylation* relates the ratio of load and phosphorylation conductance to the degree of coupling.[1] Thus, for any given degree of coupling and value of the conductance of phosphorylation, we can exactly calculate the value of the load conductance which guarantees the operation of oxidative phosphorylation at optimal efficiency.

In what follows, two problems associated with the condition of conductance matching shall be discussed. First of all, we note that q appears as a free parameter in relation (9). Therefore, the first question which comes to the mind is: What is an appropriate degree of coupling of oxidative phosphorylation in the living cell? Suppose that we have found such a value and further suppose that this degree of coupling as well as the conductance of phosphorylation are constant during short time intervals since they can be regarded as being built into the machinery of oxidative phosphorylation. Then in order to fulfill conductance matching, also the load conductance should be constant. This, however, is very unlikely to be the case in a living cell. Hence, we are left with the unpleasant but inescapable consequence that a nonconstant fluctuating load conductance would compromise the condition of conductance matching and that therefore the system could hardly ever operate at optimal efficiency. Hence, a device is sought by which this unwanted effect of a fluctuating load on efficiency could be minimized. The detailed investigation of this problem will lead to the concept of thermodynamic buffering.

II. THE DEGREE OF COUPLING OF OXIDATIVE PHOSPHORYLATION

Let us first consider the problem of an appropriate degree of coupling of oxidative phosphorylation in the cell. The solution to this question depends entirely on what output function is to be optimized. We might for example require a maximal net flow of ATP at optimal efficiency $(J_p)_{opt}$. As is evident from Fig. 4 there is a unique degree of coupling q_f, which corresponds to the maximum of this output function (see also Table I). Such a low value of the degree of coupling has never been experi-

TABLE I.   Physiologically Meaningful Values of the Degree of Coupling[a]

n	Physical meaning of $f(\alpha)$	Degree of coupling	Constant
1	$(J_p)_{opt}$	$q_f = 0.786$	ZL_0X_0
2	$(J_pX_p)_{opt}$	$q_p = 0.910$	$L_0X_0^2$
3	$(J_p\eta)_{opt}$	$q_f^{ec} = 0.953$	ZL_0X_0
4	$(J_pX_p\eta)_{opt}$	$q_p^{ec} = 0.972$	$L_0X_0^2$

[a] General output function: $f(\alpha) = tg^n (\alpha/2) \cos \alpha \cdot \text{constant}$ with $\alpha = \sin^{-1} (q)$. List of q's maximizing the general output function for integer exponents n from 1 to 4. For further details see text.

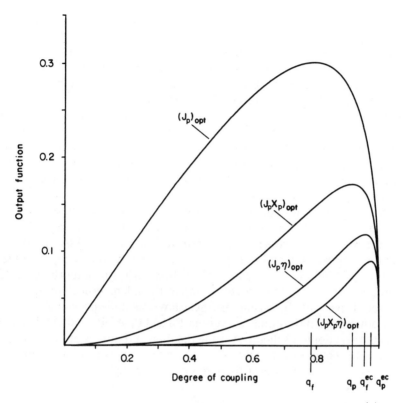

Fig. 4.   Mitochondrial output functions. Plot of the output function $f(\alpha) = tg^n \left(\dfrac{\alpha}{2}\right) \cos \alpha$ (Table I) as a function of the degree of coupling q for integer values of the exponent $n = 1, 2, 3, 4$.

mentally observed in oxidative phosphorylation. However, Lahav, Essig, and Caplan[4] have measured a degree of coupling very close to q_f in the Na^+ transport system in frog skin. This preparation, like other epithelia, is well known to have a very active Na^+ transport system. Hence there seem to be examples where a maximal flow situation is of biological relevance.

Instead of maximizing the net flow of ATP we might require to maximize the output power of oxidative phosphorylation at optimal efficiency $(J_p X_p)_{opt}$. Figure 4 shows that there is a unique degree of coupling q_p which maximizes this output function (Table I). As yet, there is no experimental system available which would tell whether the degree of coupling q_p is of any biological interest.

In the two previous optimization problems no attention was payed to the energy costs for obtaining a maximal output power. It appears, however, natural to assume that the limited availability of energy in a living cell imposes an additional and important constraint on the above optimizations because the living cell is forced to use the available reducing equivalents of the substrates in the most economic way. A convenient measure for the energy costs is the efficiency η which we may consider as a weight acting on the two output functions considered previously.

Figure 4 shows that the economic net rate of ATP synthesis $(J_p\eta)_{opt}$ as well as the economic output power $(J_pX_p\eta)_{opt}$ at optimal efficiency both assume a maximum at the unique values of the economic degrees of coupling q_f^{ec} and q_p^{ec}, respectively (Table I).

The experimental demonstration for the relevance of these economic degrees of coupling for oxidative phosphorylation in the living cell is obtained from livers perfused with substrate free media.[5] Assume that the reaction scheme depicted in Fig. 1 supplemented by the adenylate kinase reaction (see below) represents the reactions of major importance determining the steady-state concentrations of the adenine-nucleotides in the cytosol of the liver. By assuming linear laws, as done previously, we can then obtain the differential equations for ATP, ADP, and AMP on the basis of this adopted scheme. Setting the time derivatives of these adenine nucleotides equal to zero allows to calculate the steady state solutions of this system as a function of the degree of coupling[6]

$$\text{ATP}_{ss} = \frac{e^{\psi}\Sigma}{1 + e^{\phi} + e^{\psi}} \tag{10}$$

$$\text{AMP}_{ss} = \frac{e^{\phi}(\Sigma - \text{ATP}_{ss})}{1 + e^{\phi}} \tag{11}$$

$$\text{ADP}_{ss} = \Sigma - \text{ATP}_{ss} - \text{AMP}_{ss} \tag{12}$$

where $\Sigma = \text{ATP} + \text{ADP} + \text{AMP}$ represents the conservation of the adenine nucleotides in the cytosol. The exponents ϕ and ψ are

$$\phi = \frac{(\Delta G^{\circ\prime}_{phos} - \Delta G^{\circ\prime}_{AK} - RT \ln P_i)(1 + \theta\sqrt{1 - q^2}) - qX_0/Z}{(1 + \theta\sqrt{1 - q^2})RT} \tag{13}$$

$$\psi = \frac{(RT \ln P_i - \Delta G^{\circ\prime}_{phos})(1 + \theta\sqrt{1 - q^2}) + qX_0/Z}{(1 + \theta\sqrt{1 - q^2})RT} \tag{14}$$

$\Delta G^{\circ\prime}_{phos}$ and $\Delta G^{\circ\prime}_{AK}$ are the standard free energies of phosphorylation and of the adenylate kinase reaction, respectively, at pH 7. P_i is the phosphate concentration in the cytosol and θ is a parameter measuring deviation from conductance matching. All other symbols have the usual meanings. Figure 5 shows plots of the cytosolic adenine nucleotide concentrations as a function of the degree of coupling in the range $0.9 < q < 1$.

On these theoretical curves two sets of experimental data are inserted: the cytosolic adenine nucleotide concentrations measured in livers from fed rats (F) and starved rats (H).[5] In the livers of the fed rats the degree of coupling is astonishingly close to q_p^{ec} (Table I). These livers have been regarded to be in a "metabolic resting state."[7] Hence, for the metabolic resting state, the maximization of the economic output power seems to be of importance. In contrast, in the livers from starved rats the degree

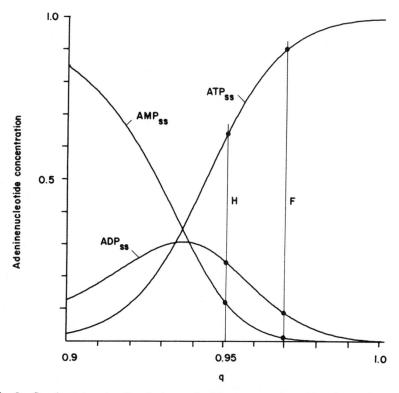

Fig. 5.   Steady-state cytosolic adenine nucleotide concentrations. Plot of equations (10)–(14) as a function of the degree of coupling in the interval $q\epsilon(0.9,1)$. Values of the parameters $\Delta G^{0'}_{phos} = 8.5$ kcal/mole, $\Delta G^{0'}_{AK} = 0.15$ kcal/mole, $P_i = 0.008\ M$, $X_o = 50$ kcal/mole, $Z = 3$, $\theta = 1$. Inserted points: experimental values from perfused livers.[5] Normalized plots with $\Sigma = 1$.

of coupling very closely corresponds to q_f^{ec} (see also Table I). In a starved animal with depleted liver glycogen stores, a major metabolic task of the liver consists in supplying the organism with glucose in the absence of external sources of this substrate. The reaction sequence responsible for the synthesis of glucose from endogenous low carbon substrates is gluconeogenesis.[8] This metabolic pathway is subject to hormonal control. Since gluconeogenesis utilizes ATP to drive the endergonic synthesis of glucose it appears therefore adequate that in livers from starved rats the degree of coupling of oxidative phosphorylation corresponds to q_f^{ec} which clearly favors a maximal economic net flow of ATP.

It is important to note that this satisfactory fit of experimental and theoretical data required $\theta = 1$. Hence the conditions of conductance matching was fulfilled on a time average in both situations considered, namely, livers from fed as well as from starved rats. In summary, it may be concluded that both q_p^{ec} and q_f^{ec} appear to be relevant for the operation of oxidative phosphorylation in the liver cell depending on the special output requirements, which favor either one or the other of these economic degrees of coupling as predicted by theory.[1] The selection of the appropriate degree of coupling in combination with a matched load conductance thus allows oxidative phosphorylation to adapt in the most economic way to the characteristic energy demands of different metabolic conditions.

III. THERMODYNAMIC BUFFERING

After having obtained some general ideas about appropriate degrees of coupling let us now investigate how nature could cope with fluctuating load conductances, which would endanger the conditions of conductance matching in the living cell. The experiment shown in Fig. 6a has led to the discovery of a new bioenergetic regulatory principle, which allows oxidative phosphorylation to operate at optimal efficiency in the presence of mismatched loads. In this experiment, isolated rat liver mitochondria were incubated in the presence of increasing hexokinase concentrations. Hexokinase catalyzes the ATP utilizing formation of glucose 6-phosphate from glucose and thus represents a load imposed on the machinery of oxidative phosphorylation.[6] As expected, an increase in the hexokinase concentration led to a drop of the phosphate potential. Since X_0 was held constant this decreasing phosphate potential resulted in a concomitant drop of the force ratio toward level flow. As theoretically predicted the efficiency increased by shifting the force ratio toward level flow. However, once the system reached the state of optimal efficiency it was not possible to drive this energy converter past this state by further increasing the hexokinase concentration in the incubation medium. This striking experimental result suggested a reaction occurring in the system which was able to prevent the drop of the phosphate potential below the value permitting optimal efficiency. The measurement of the adenine nucleotide concentrations in the incubation medium showed a pronounced accumulation of AMP in the presence of high hexokinase concentrations.[6] The well-known reaction giving rise to AMP formation in this system is catalyzed by adenylate kinase (Fig. 7). This enzyme is localized in the intermembrane space of the mitochondrial preparation.

The simple reaction scheme depicted in Fig. 7 allows an intuitive understanding of the observed experimental result. An increasing load conductance results in a drop of the steady-state ATP concentration and in an increase of the steady-state ADP concentration. This shift leads to a decrease of the phosphate potential below the value permitting optimal efficiency. Through the adenylate kinase reaction, however, part of the ADP is reconverted into ATP and into AMP which accumulates. Therefore, this reaction counteracts the effect of too high load conductances on the phosphate potential. Since the adenylate kinase reaction is readily reversible the opposite effect on the phosphate potential, namely, a lowering of this potential toward the value permitting optimal efficiency, is to be expected in the presence of too low load conductances, provided of course that AMP be present in the system. From this reasoning it appears that the adenylate kinase reaction is able to buffer the phosphate potential near the value permitting optimal efficiency of oxidative phosphorylation in the presence of mismatched loads. Note that here for the first time an enzyme is proposed to be involved in a regulatory mechanism acting on a thermodynamic potential. Since every reversible ATP utilizing reaction can in principle be expected to exhibit a similar effect we can generalize the above observation to a new bioenergetic regulatory principle which we call *thermodynamic buffering*.[6] Thus, for example creatine kinase and arginine kinase are likely candidates to join the family of the thermodynamic buffer enzymes.

After these general comments let us further test the idea of thermodynamic buffering on an experimental basis by repeating the above experiment but this time in the presence of an inhibitor of adenylate kinase, namely, diadenosine pentaphosphate. As is depicted in Fig. 6b the buffering effect of the adenylate kinase is abolished by inhibiting this enzyme and it becomes now possible to drive the system beyond the state of optimal efficiency by increasing the hexokinase concentration in the medium. Note that it was not possible to measure points closer to level flow than the ones shown in the figure. This is due to technical reasons. At the lowest phosphate potentials the ATP/ADP ratios where of the order of 10^{-3} which represents the limit of what can be accurately measured with enzymatic assays by using dual wavelength spectrophotometry.

In order to obtain a more intuitive insight into the mechanism of thermodynamic buffering we calculated the effects of thermodynamic buffering on the entropy production of the system. The entropy production of oxidative phosphorylation with an attached load is given in equation (8). A convenient way to introduce the contribution of the adenylate kinase reaction to this system is to consider L_l as an overall load conductance embracing the effects of the adenylate kinase reaction as well as the effects of the true extrinsic load conductance of the irreversible ATP utilizing

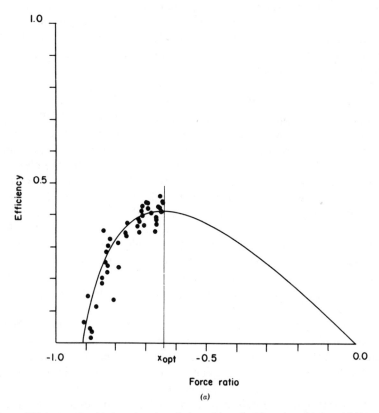

Fig. 6. Efficiency of oxidative phsophorylation with and without hexokinase. (a) From the experimentally determined values of J_p, J_o, X_p, and X_o, η was calculated according to equation (5) and plotted as a function of the force ratio. A phenomenological stoichiometry of $Z = 2.83$ was used. The solid line represents the theoretical curve for $q = 0.94$. The data of three pooled experiments carried out exactly as described in Fig. 2 are plotted.

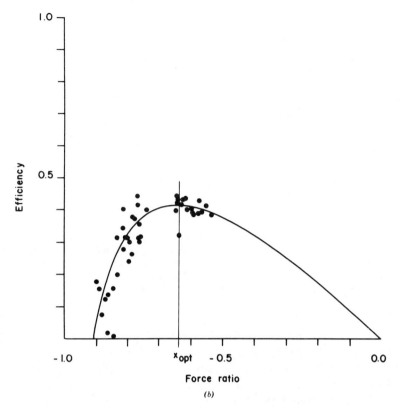

Fig. 6. (*continued*) (*b*) Pooled data from three experiments carried out as in Fig. 2 but containing 20 μM P^1,P^5-di(adenosine-5'-)pentaphosphate. For further details see ref. 6.

reactions in the cell, designated hereafter as L_l^e. This overall load conductance is[6]

$$L_l = L_{AK} \left[\frac{\alpha + RT \ln(1 + e^{-(\beta + X_P)/RT})}{X_P} - 1 \right]^2 + L_l^e \qquad (15)$$

where $\beta = \Delta G_{phos}^{\circ\prime} - RT \ln P_i$ and $\alpha = \Delta G_{AK}^{\circ\prime} + RT \ln[M/(1 - M)] - \beta$. M represents the AMP concentration normalized by $M = AMP/\Sigma$. L_{AK} is the conductance of the adenylate kinase reaction.

Two situations can now be distinguished: either $L_{AK} = 0$, then we have the entropy production of the unbuffered case; or $L_{AK} \neq 0$, then we have the entropy production of the buffered case. To clearly distinguish between these two situations let us call the entropy production of the un-

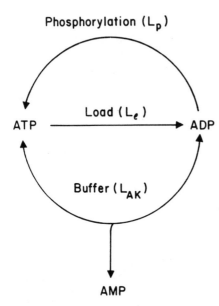

Phosphorylation (L$_p$)

ATP ——— Load (L$_\ell$) ———→ ADP

Buffer (L$_{AK}$)

AMP

Fig. 7. Reaction scheme for oxidative phosphorylation with load and adenylate kinase. L_p, conductance of phosphorylation; L_l, load conductance; and L_{AK} conductance of adenylate kinase reaction. For details see text.

buffered case $\dot{S}_U$ and the buffered case $\dot{S}_B$. Figure 8 shows plots of $\dot{S}_U$ (panel a) and $\dot{S}_B$ (panel b) as a function of the force ratio for various values of the load conductance. As is evident from Fig. 8a a variation of the load conductance leads to a large variation of the steady-state force ratio about the value permitting optimal efficiency x_{opt}. Only at the matched value of the load conductance the steady-state value of the force ratio is exactly equal to x_{opt}. On the other hand, the plot of $\dot{S}_B$ in Fig. 8b shows that the variation of the steady-state force ratios about x_{opt} is much smaller for the same variations of the load as in Fig. 8a as a result of thermodynamic buffering.

Finally, we might wish to observe thermodynamic buffering at work in an intact cell in a fluctuating environment. Unfortunately there is no experimental technique available at present which would permit a continuous measurement of the phosphate potential in the cytosol of a living cell. Therefore, we are left with the only possibility to give such a demonstration by a numerical computer simulation.

For this simulation we took again the model as was used for the previous calculations of the steady state adenine nucleotides, namely, oxidative phosphorylation with an attached load plus the adenylate kinase reaction. The only modification introduced into this scheme now was to consider a fluctuating rather than a constant load conductance. In order to arrive at a realistic description, a stationary process with a Lorentzian type of power spectrum was chosen to perturb the matched mean value of the load conductance $\langle L_l^e \rangle$

$$L_l^e(t) = \langle L_l^e \rangle + \zeta(t) \tag{16}$$

whereby

$$\dot{\zeta}(t) = -\tau\zeta(t) + \xi(t) \tag{17}$$

describes the stationary Ornstein–Uhlenbeck process $\zeta(t)$ (colored or real noise) represented by the Langevin equation.[9] τ is the inverse correlation time; $\xi(t)$ is the white noise with variance σ.

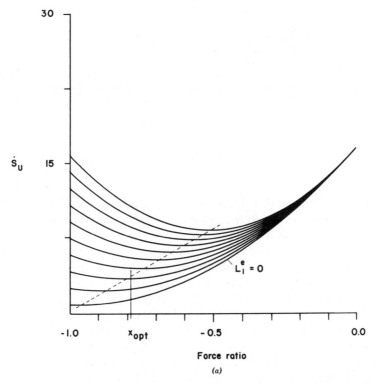

Fig. 8.   Effect of thermodynamic buffering on entropy production. (a) Plot of $\dot{S}_u$ versus x [equations (8) and (15) with $L_{AK} = 0$]. The plots were normalized with $L_p = L_o = X_2 = 1$. The other values in the plot: $\Delta G_{phos}^{0'} = 8.5$ kcal/mole, $\Delta G_{AK}^{0'} = 0.15$ kcal/mole, $P_i = 0.008\ M$, and $M = 0.005$. L_l^e was varied between 0.0 and 0.9 in steps of 0.1. Thin line, x_{opt} for q_p^{ec}; broken line, loci of steady states.

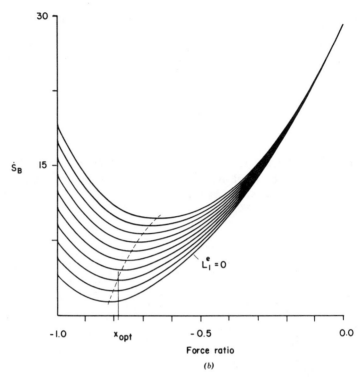

Fig. 8. (*continued*) (*b*) plot of $\dot{S}_B$ versus x [equations (8) and (15) with $L_{AK} = 1$].

The main result of this simulation is depicted in Fig. 9. Figure 9a shows the fluctuations of x about the value x_{opt} as a consequence of a fluctuating load conductance in the absence of adenylate kinase. Due to the shape of the efficiency function (Fig. 3) these deviations from x_{opt} resulted in deviations from η_{opt} in the unbuffered case. It might appear here that the deviations from optimal efficiency are only tiny. But it must be realized that these tiny deviations integrated during a lifetime result in a terrible net loss of ATP. In the simulation shown in Fig. 9b exactly the same time history of fluctuations was chosen as in the previous simulation but this time in the presence of the adenylate kinase reaction. From the comparison of the two simulations it readily appears that the fluctuations of the force ratio are largely damped out and that only the low frequency perturbations survive. As a consequence of this smoothing of x near x_{opt} the system can now operate always close to optimal efficiency.

It is interesting to note that the subcellular distribution of the thermodynamic buffer enzymes is also in accord with their physiological role proposed here. Adenylate kinase as well as creatine kinase are exclusively localized on the outside of the mitochondrial matrix whereas oxidative phosphorylation is confined to the matrix space of the mitochondria.[10] This special compartmentation of buffer reaction and energy converter should allow these enzymes to act like a filter. After the perturbation of the phosphate potential by a mismatched load in the cytosol, the buffer first establishes the phosphate potential permitting optimal efficiency before the ADP is handed over to the adenine nucleotide translocase which transports the ADP into the matrix in exchange with ATP. Thus, the matrix space should only see the arrival of a tuned phosphate

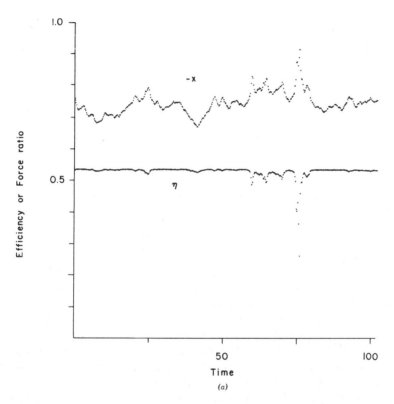

Fig. 9.   Oxidative phosphorylation with a fluctuating load. (a) Plot of efficiency and force ratio versus time in the absence of adenylate kinase: $L_{AK} = 0$. The other parameters used for the simulation were as follows. Initial conditions: ATP = 1.03, ADP = 0.234, AMP = 0.738, $\xi = 11\ N(0,\ \sigma^2/2\tau)$. Constants: $L_p = 0.247$, $L_{po} = 0.0785$, $L_f^s = 0.0749$, $L_o = 0.0274$, $\Delta G_{phos}^{0'} = 8.5$ kcal/mole, $P_i = 0.008\ M$, $\Delta G_{AK}^{0'} = 0.15$ kcal, $\tau = 20$, $\sigma = 0.1$, time step of integration $h = 0.01$ time units. For further details see text and ref. 6.

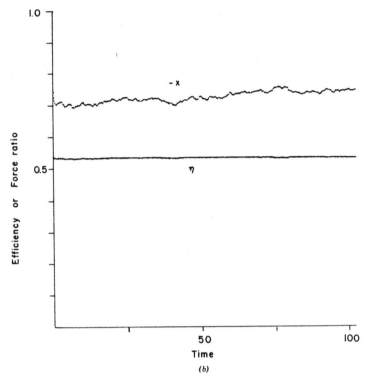

Fig. 9. (*continued*) (*b*) Same plot as in 9 (a) with adenylate kinase: $L_{AK} = 5.0$.

potential allowing to perform the phosphorylation of ADP at optimal efficiency.

In summary, the living cell has selected special reversible ATP-utilizing reactions that permit oxidative phosphorylation to operate at optimal efficiency in the presence of a fluctuating load conductance. Therefore, the thermodynamic buffering is not a principle of optimization proper to the process of oxidative phosphorylation itself. It is rather an additional reaction attached to oxidative phosphorylation which is not only chemically but also spatially separated from the energy converting process. In this respect, the optimization of efficiency through the choice of a proper degree of coupling is very different from the optimization through thermodynamic buffering. It remains to be seen whether the thermodynamic buffer enzymes are of a vital importance for the living cell. The ubiquitous distribution of adenylate kinase in all cells having oxidative phosphorylation would indicate at least that thermodynamic buffering plays a very important physiological role.

IV. A NONLINEAR ENERGY CONVERTER

Let us finally consider an entirely different possibility for the optimization of oxidative phosphorylation. The experiment presented at the outset (Fig. 2) revealed linear and symmetric relations between flows and forces in a far-from-equilibrium regime. This experimental observation is very astonishing by itself since it appears to be in sharp contrast with the well-known result that the linear domain of nonequilibrium thermodynamics is severely limited to the immediate vicinity of equilibrium through the inequality $| X_i | \ll RT$.[3] In oxidative phosphorylation, however, we typically deal with phosphate potentials around 9–15 kcal/mole and redox potentials of the substrates in the order of 40–50 kcal/mole. In the light of this large discrepancy, the linear and symmetric behavior of oxidative phosphorylation therefore appears rather to be the result of an artificial and sophisticated regulation mechanism maintaining linearity and symmetry far from equilibrium, than to be a natural consequence of a near-equilibrium regime. The existence of a sophisticated regulation of oxidative phosphorylation is also indicated by the complex assembly of the enzymes involved in this process in the membrane. Thus, for example, the enzyme involved in the final step of electron transport, the cytochrome c oxidase, consists of seven subunits out of which only two appear to be catalytic subunits and the other five are presently regarded as regulatory subunits. Another example is the ATPase responsible for the phosphorylation of ADP to ATP. This enzyme consists of five complexes with a total of eight subunits and only the "β" complex appears to be catalytic whereas the other complexes are believed to be involved in regulation. (For a recent review see for example Azzi and Casey.[11])

These considerations lead inevitably to the crucial question: Is there any advantage for the performance of oxidative phosphorylation gained through linear and symmetric relations between flows and forces? At present it is impossible to give a final answer to this question. However, the results of a few first steps toward this answer will be briefly outlined.

Thus far, we studied the process of oxidative phosphorylation on a purely phenomenological basis. Our phenomenological scheme proved to contain enough information to allow the uncovering of the optimization principles discussed earlier. In order to give an answer to the question whether linearity and symmetry are advantageous for oxidative phosphorylation we have to partly abandon this purely phenomenological analysis and to make a step toward a mechanistic description of the system. This is tantamount to partially reticulating the box representing oxidative phosphorylation in Fig. 1. Following the generally accepted view of a chemiosmotic type of coupling mechanism of oxidative phosphorylation[12] we can envisage this process to consist of two energy converters operating in series as depicted in Fig. 10. In fact, these energy converters are both active transport processes interlinked through a common transported ionic species, the H^+ ion. In a first step, a redox driven H^+ pump translocates H^+ ions from the matrix space to the cytosol, (J_H), thereby establishing an electrochemical potential difference of H^+ ions, X_H, across the inner membrane. In a second step, this electrochemical H^+ gradient is used to drive the ATPase H^+ pump backward, which results in ATP formation. This scheme, depicted in Fig. 10, seduces us to interpret the redox driven H^+ pump as the energy converter establishing the electrochemical force X_H, whereas the ATPase H^+ pump then merely appears as an energy transducer that scales this force X_H into the output force of oxidative phosphorylation X_p. Thus, in order to give an answer to the question of possible advantages of linearity and symmetry we focus our attention to the redox H^+ pump. The justification of this choice comes also from the recent experimental observation that the redox H^+ pump in rat liver mitochondria shows linear as well as symmetric relations between flows (J_0 and J_H) and the output force X_H at a constant input force X_0.[13]

Since the mechanistic details of the redox H^+ pump are unknown we may consider a hypothetical H^+ channel which can bind and release H^+ ions on either side of the membrane and where the oxidation of the substrate is compulsorily linked to a reorientation of the opening of the channel through a gating mechanism, for example (Fig. 11). This simple scheme can be interpreted as a general coarse grained picture of the mitochondrial H^+ pump. In addition to being simple and general this scheme allows us to use the diagrammatic method of T. L. Hill to obtain the general relations between flows and forces for this energy converter.[14] The advantage of Hill's method is that it permits to combine thermodynamic with kinetic

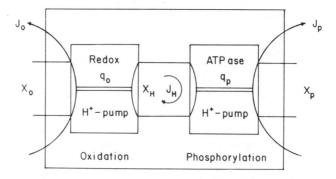

Fig. 10.   Operational scheme of oxidative phosphorylation. For further details see text.

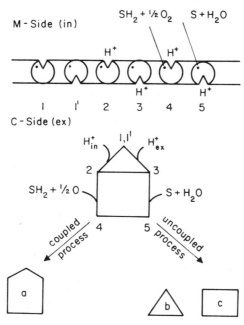

Fig. 11.   Diagrammatic description of mitochondrial redox H^+ pump. For further details see text.

considerations in a convenient and elegant way. The different states of the H^+ channel can be mapped onto the nodes of a graph in the way depicted in Fig. 11. In this graph there are three cycle graphs possible which permit a steady-state operation of the pump. As appears from Fig. 11 the cycle a stands for the transport of H^+ ions coupled to the redox reaction whereas cycles b and c stand for the uncoupled contributions to H^+ transport and redox reaction, respectively. Following the rules given in Hill's book one obtains the general nonlinear relations between flows and forces in an algorithmic fashion from the individual cycle graphs. For the net H^+ transport we obtain from cycles a and b[15]

$$J_H^{nl} = n[(a + b)(e^{nX_H} - 1) + a(e^{X_0} - 1) + a(e^{X_0} - 1)(e^{nX_H} - 1)]$$

(18)

and for the net oxygen consumption in a similar manner from cycles a and c

$$J_0^{nl} = a(e^{nX_H} - 1) + [a + c'(1 + \gamma)](e^{X_0} - 1)$$
$$+ (a + c')(e^{nH_H} - 1)(e^{X_0} - 1) \quad (19)$$

where n is the mechanistic stoichiometry of the H^+ pump; a, b, and c' are coefficients consisting of rate constants describing the transitions among the different states in cycles a, b, and c, respectively; and γ stands for the ratio of dissociation constants for H^+ ions on both sides of the membrane. Note that in our description the forces are to be understood as expressed in RT units. In the limit of small forces these equations yield the linear relations

$$J_H^l = n^2(a + b)X_H + anX_0 \qquad (20)$$

$$J_0^l = anX_H + [a + c'(1 + \gamma)]X_0 \qquad (21)$$

which have the same structure as our phenomenological equations (1) and (2). From the usual definitions of q and Z [equations (3) and (4)] we obtain now an intuitive insight into the meaning of these parameters on a mechanistic level

$$q_0 = \frac{a}{\sqrt{(a + b)[a + c'(1 + \gamma)]}} \qquad (22)$$

$$Z_0 = n\sqrt{\frac{a + b}{a + c'(1 + \gamma)}}$$

From these relations it readily emerges that the cycles b and c introduce a slip into the pump such that $q_0 < 1$ if b and c' do not vanish simultaneously. Furthermore, the phenomenological stoichiometry Z_0 is not equal to the mechanistic stoichiometry n unless $q = 1$, that is, $b = c' = 0$, although contrary claims have been made in the literature.[16] From recent experimental results obtained with isolated rat liver mitochondria it can be estimated that in the redox H^+ pump the contributions of the b and c cycle are about 3.5 and 2% of a cycle, respectively.[13] With equations (18) and (19) and equations (20) and (21) we have now obtained two sets of equations: the equations describing the nonlinear energy converter and the equations describing the linear energy converter. By virtue of the definition of q_0 and Z_0 common to both converters, these two types of converters can now be compared in a physically consistent manner. It cannot be overstressed that by making this comparison in what follows below we will, in fact, consider two different objects. We do not interpret the linear energy converter to be the result of a small force approximation of the nonlinear converter but, instead, to be a feedback regulated device being able to operate in a linear fashion far from equilibrium.

Figure 12 depicts the dependence of the efficiency of the nonlinear converter

$$\eta^{nl} = -\frac{J_H^{nl} X_H}{J_0^{nl} X_0} \tag{23}$$

as a function of the force ratio on the magnitude of the input force X_0. For small values of X_0 the efficiency of the nonlinear converter is very close to the efficiency of the linear converter

$$\eta^{l} = -\frac{J_H^{l} X_H}{J_0^{l} X_0} \tag{24}$$

As the input force increases toward the physiologically relevant range of the forces, however, η^{nl} drops to small values and the force ratio permitting optimal efficiency is shifted toward level flow. This behavior of

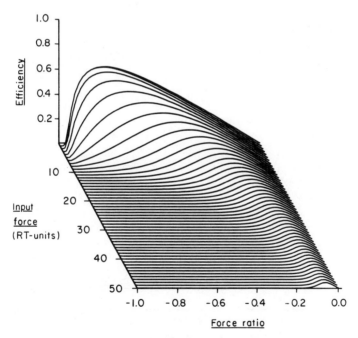

Fig. 12.   Dependence of efficiency of a nonlinear energy converter on the input force. Plot of equations (18),(19), and (24) as a function of X_H/X_0 for various values of X_0 $\epsilon(0,50)RT$ units. Normalized by $b = c'(1 + \gamma)$ and $\gamma = 1$. For further details see text.

the nonlinear converter is radically different from the behavior of the linear one since in the latter case η^l_{opt} as well as x^l_{opt} are invariant with respect to the input force X_0 [see equations (6) and (7)].

This different behavior of the two converters has two consequences which are of an immediate biological interest. As depicted in Fig. 13 the nonlinear energy converter can only maintain very small output forces (~1 kcal/mole) at optimal efficiency at the physiological values of the input force. These small output forces, which are near equilibrium, could hardly ever be sufficient for the onset and maintenance of dissipative structures in the living cell.[3] In contrast, the linear energy converter can easily maintain high output forces at optimal efficiency.

An even more dramatic effect is observed when comparing the efficiencies of the nonlinear and the linear energy converter. As shown in Fig. 14 the separatrix $\eta^{nl}/\eta^l = 1$, which delimits the regions where the nonlinear and the linear energy converter are more efficient, declines

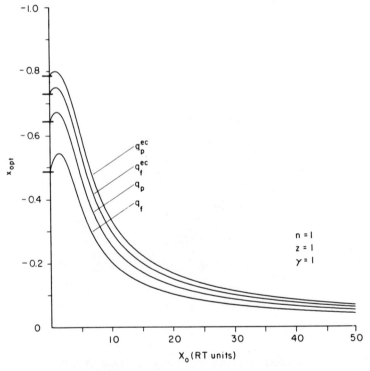

Fig. 13.   Dependence of optimal force ratio on input force. The force ratio x_{opt} for optimal efficiency states in the previous figures were plotted as a function of X_0 for the values of the degrees of coupling indicated in the figure. Normalized by $b = c'(1 + \gamma)$ and $\gamma = 1$.

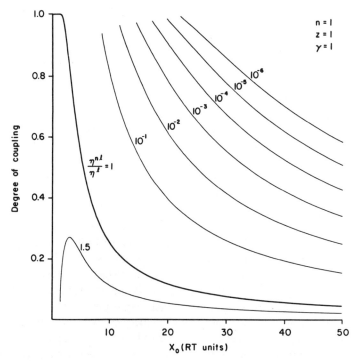

Fig. 14.   Optimization through linearity. Various values of the ratio η^{nl}/η^{l} [equations (23) and (24)] plotted as a function of q and X_0. Normalized by $b = c'(1 + \gamma)$ and $\gamma = 1$.

already at small values of the input force to unphysiologically low values of q. In the physiologically meaningful range of values for X_0 and q, however, the linear energy coverter is at least 100,000 times more efficient than the nonlinear one. Thus, the efficiency gained through linearity is definitely not only a matter of a few percent! It can be expected that such a dramatic gain in efficiency would result in a very strong evolutionary drive toward linearity of the H$^+$ pump. Preliminary calculations show that these properties of the redox H$^+$ pump carry over to the whole process of oxidative phosphorylation, provided that the ATPase H$^+$ pump is tightly coupled. Experimental evidence shows that exactly this is the case.[13]

These considerations explain the tremendous advantage of having linear relations between flows and forces. But what about symmetry? The model in Fig. 11 permits no answer to this question because as is apparent from equations (20) and (21) symmetry is already built into that model. At present it is therefore not clear whether the experimentally observed symmetry is a mere coincidence, or an inescapable consequence of the underlying mechanism, or the result of an as yet undiscovered additional optimizing principle. This problem has to remain open until a general argument can be put forward to support one of these alternatives.

V. DISCUSSION

In summary, it appears now that the efficiency of far-from-equilibrium energy converters is subject to a new kind of optimizing principle in addition to the ones encountered so far. This principle might be called *optimization through linearity*. It is important to stress again the fundamental difference between the kind of linearity observed near equilibrium and far from equilibrium. Albeit both kinds of linearity obey the very same form of the classical Onsager relations, the former kind is a natural consequence of a near equilibrium regime whereas the latter appears to be the result of an efficiency optimizing principle of biological energy converters far from equilibrium. This observation can perhaps be paraphrased by saying that the living cell has managed to prolong a linear domain of nonequilibrium with the aid of enzyme catalyzed and regulated reactions. It remains to be seen how large this linear region is. The presently available experimental evidence indicates linearity between flows and forces within the range of the forces considered to be of physiological relevance.

There remain many other important problems to be solved in the future. First of all, it is very likely that additional optimizing principles to the ones considered here are of importance for the optimization of biological energy conversions. One of these additional principles has recently been discovered by Richter and Ross.[17] Their principle is related to the temporal behavior of completely coupled systems such as glycolysis and has therefore to be understood on a basis of frequency matching rather than on a basis of conductance matching as considered here. It seems not impossible that such a principle could also be of importance in mitochondrial energy conversion, especially in situations characterized by metabolic oscillations. This brings us to a final remark. So far we have looked at oxidative phosphorylation in splendid isolation. This analysis should, however, be extended in order to understand more clearly how the process of oxidative phosphorylation is interwoven into the complex fabric of intermediary metabolism in the living cell.

Acknowledgments

This work was supported by grants from the Swiss National Science Foundation. The author would like to express his gratitude for stimulating discussions to Professors S. R. Caplan, Agnessa Babloyantz, G. Nicolis, and I. Prigogine. The expert technical assistance of Miss Lilly Lehmann and Margret Over is gratefully acknowledged.

References

1. J. W. Stucki, *Eur. J. Biochem.*, **109**, 269–283 (1980).
2. O. Kedem and S. R. Caplan, *Trans. Faraday Soc.*, **21**, 1897–1911 (1965).
3. G. Nicolis and I. Prigogine, *Self-Organization in Nonequilibrium Systems*, Wiley, New York, 1977.
4. J. Lahav, A. Essig, and S. R. Caplan, *Biochim. Biophys. Acta*, **448**, 389–392 (1976).
5. S. Soboll, R. Scholz, and H. W. Heldt, *Eur. J. Biochem.*, **87**, 377–390 (1978).
6. J. W. Stucki, *Eur. J. Biochem.*, **109**, 257–267 (1980).
7. S. Soboll, Doctoral thesis, University of Munich, 1977.
8. M. C. Scrutton and M. F. Utter, *Annu. Rev. Biochem.*, **37**, 249–302 (1968).
9. M. C. Wang and G. E. Uhlenbeck, *Rev. Med. Phys.*, **17**, 323–342 (1945).
10. G. L. Sottocasa, B. Kuylenstierna, L. Ernster, and A. Bergstrand, *Meth. Enzymol.*, **10**, 448–463 (1967).
11. A. Azzi and R. P. Casey, *Meth. Cell Biol.*, **28**, 169–185 (1979).
12. P. Mitchell, *Chemiosmotic Coupling and Energy Transduction*, Glynn Res. Ltd., Bodmin, Cornwall, England, 1968.
13. D. Pietrobon, M. Zoratti, G. F. Azzone, J. W. Stucki, and D. Walz, *Eur. J. Biochem.*, **127**, 483–494 (1982).
14. T. L. Hill, *Free Energy Transduction in Biology*, Academic Press, New York, 1977.
15. J. W. Stucki, M. Compiani, and S. R. Caplan, *Biophysical Chemistry*, **18**, 101–109 (1983).
16. H. Rottenberg, *Biochim. Biophys. Acta*, **549**, 225–253 (1979).
17. P. H. Richter and J. Ross, *J. Chem. Phys.*, **69**, 5521–5531 (1978).

COMMENTARY: DISSIPATION REGULATION IN OSCILLATORY REACTIONS. APPLICATION TO GLYCOLYSIS

JOHN ROSS and PETER H. RICHTER

Department of Chemistry, Stanford University, Stanford, California

We report briefly on some recent work on efficiency of irreversible processes and an analysis of glycolysis. The operation of thermal and chemical engines with nonzero power output requires a minimum set of irreversible processes.[1-7] We have studied the behavior of oscillatory chemical reactions in the working fluid of a chemical engine and find that the efficiency of the engine, defined as the ratio of output to input power, shows interesting resonance responses.[6,8] When the period of the oscillatory reaction is about the same as the engine cycle time (or a simple multiple thereof), then the efficiency (or equivalently, the dissipation) depends strongly on the relative phase between the oscillations on the input and the output side of the engine. For properly adjusted phase, the efficiency is substantially increased (the dissipation decreased) near resonances.

We have applied the concept of regulation of dissipation of resonance response (and phase control) to a study of glycolysis. Measurements on concentrations of intermediates (from fructose 6-phosphate to pyruvate) show oscillations of chemical concentrations. The Gibbs free energy profile of the reaction mechanism, obtained by measurement of concentrations and phase lags of the oscillation of one intermediate with respect to a prior intermediate, is highly structured. The PFK reaction has a substantial Gibbs free energy change (ΔG), and is the primary oscillophor. The PK reaction also has a large ΔG, and has a mechanism that may lead to oscillations. We suggest that the PFK reaction drives the PK reaction with a resonance frequency, the phase being adjusted by the GAPDH reaction so as to achieve dissipation regulation. We estimate that a reduction of 5–10% in entropy production may thus be achieved. The possibility of dissipation reduction or, more generally, its regulation has obvious evolutionary advantages and may for that reason have developed in glycolysis.

References

1. F. L. Curzon and B. Ahlborn, *Am. J. Phys.*, **43**, 22 (1975).
2. B. Andresen, R. S. Berry, A. Nitzan, and P. Salamon, *Phys. Rev. A*, **15**, 2086 (1977).
3. P. Salamon, B. Andresen, and R. S. Berry, *Phys. Rev. A*, **15**, 2094 (1977).
4. B. Andresen, P. Salamon, and R. S. Berry, *J. Chem. Phys.*, **66**, 1571 (1977).
5. D. Gutkowicz-Krusin, I. Procaccia, and J. Ross, *J. Chem. Phys.*, **69**, 3898 (1978).
6. P. H. Richter and J. Ross, *J. Chem. Phys.*, **69**, 5521 (1978).
7. M. H. Rubin, *Phys. Rev. A*, **19**, 1272 and 1277 (1979).
8. P. H. Richter and J. Ross, *Biophys. Chem.*, **12**, 285 (1981).

COMMENTARY: A COMPLEX
BIOLOGICAL OSCILLATOR

E. SCHOFFENIELS

*Laboratory of General and Comparative Biochemistry,
University of Liège, Liège, Belgium*

As an extension of the contribution of Dr. Stucki, I would like to bring attention to a rather complex biological oscillator. It deals with the case of invertebrate euryhaline species, clearly a case of integration of a biological system into its environment, that is, what the biologists call adaptation.

Fitness of adaptation to the environment means that biochemical correlates of physical or chemical characteristics of the environment must exist.[1] Therefore, adaptive processes are always optimization processes be it that of metabolic sequences or of molecular properties.[2,3] To recognize that a living system is adapted to a given environment is simply to recognize its finality[4] or teleonomic characteristics.[5] Optimization being a consequence of the second law of thermodynamics, it seems to me that what Jacob or Monod call finality or teleonomy is simply the finality of the second law.

An ecosystem, like an organism, may be considered to be a collection of metabolic sequences which assume the most efficient usage of the free energy available. In this view, ecosystems and organisms alike can be considered as the expression of the natural tendency for the most efficient dissipation of free energy in a given environment. The justification of this interpretation has been given elsewhere.[6] As discussed by Prigogine and his colleagues,[7] increased free energy dissipation is paralleled by evolutionary complexity. Therefore, the thermodynamic correlates of biological evolution may be found in the tendencies toward increased energy dissipation (i.e., entropy production) and by an increase in the efficiency of energy utilization. This interpretation rests on what seems to me an unescapable conclusion, namely, that biological evolution finds its counterpart in the energetic processes of the organism.[8]

The fitness of an adaptation necessarily implies the existence of feedback loops between organism and environment that are certainly apparent

171

when considering the euryhaline crustacea. As in any other multicellular organisms, most of their energy is derived from oxidoreduction processes involving electron transport systems. It is convenient to define the energy source as being the electron donor and the energy sink is the final acceptor. The chemical nature of the electron acceptor is directly related to the physical and chemical compositions of the environment. Take the case of the chinese crab, *Eriocheir sinensis* Milne Edwards, which utilizes oxygen or α-ketonic acids as electron acceptor according to the salinity of the environment.

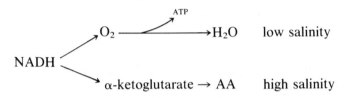

In a dilute medium, synthesis of water results with production of ATP, whereas in a concentrated medium, synthesis of amino acid occurs. In both cases the process leads to the isosmotic regulation of intracellular fluid and therefore the maintenance of a constant cell volume.[9]

Changes in the salinity of the environment are reflected in changes in the ionic composition of the cells which, in turn, affects the catalytic activity of key enzymes. As a consequence, the reducing equivalents produced are either diverted toward oxygen (low salinity) or toward α-ketoglutaric acid (high salinity).

Notice that the partial oxygen pressure remains constant and that in both situations large amplitude oscillations in NADH and some amino acid concentrations are observed.

In relation with some questions raised by Eigen and Prigogine regarding the neo-Darwinian approach to evolution I will comment further on the molecular aspects of euryhalinity. As shown in the diagramatic representation (Fig. 1), the ability for an aquatic species to invade medium of various salinities is not the result of changes at the level of *one* molecular species but changes occurring at the level of *many* enzymes. It is therefore evident that to be successful the mutations affecting one enzyme must be specified, and moreover, since more than one enzyme is involved in the process of adaptation, the mutations at the level of each of the enzymes must be correlated such that they provide the changes adequate to ensure the integration and tuning of the metabolic sequences they control. Finally, I will discuss briefly an aspect of molecular evolution that also originates in our studies of the molecular approaches to adaptation.

When considering the balance of the osmotic effectors in the cell of a

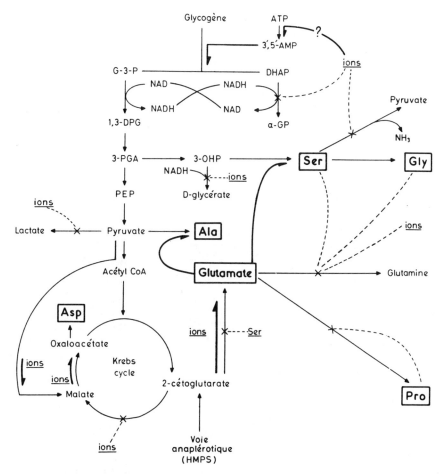

Fig. 1. Integrated scheme showing the metabolic systems of regulation operating in eu-ryhaline crustacea (after ref. 9). Broken lines indicate an inhibitory action of the effector and the heavy lines indicate activation. Notice the key role played by glutamic dehydrog-enase as well as the controls exerted on the reactions utilizing reducing equivalents. The cAMP concentration is higher in concentrated medium than in dilute medium. The hormone responsible for this effect is not yet identified. HMPS, hexoses monophosphate shunt.

euryhaline species, it is apparent that on a quantitative basis inorganic ions and organic effectors such as amino acids are the most important (Table I). Other small organic molecules that are substrates, intermediary, or end products of metabolic sequences are found in concentrations that are irrelevant when drawing an osmotic balance. It is therefore urgent to consider the mechanisms that have evolved to keep an efficient metab-

TABLE I.   Intracellular Osmotic Effectors in Muscle
of *Eriocheir sinensis* Milne Edwards Adapted to Fresh
Water or to Seawater[a]

Intracellular osmotic effectors	Concentration (mOsm/liter H_2O)	
	Fresh water	Seawater
Cl	44.6	166.9
Na	41.4	146.9
K	84.5	133.1
Ca	5.2	11.2
Mg	9.2	25.3
Subtotal	184.9	483.4
Ala	18.1	71.9
Arg	36.5	54.7
Asp (total)	3.6	11.7
Glu (total)	10.3	28.2
Gly	57.0	108.5
Ile	1.0	3.2
Leu	1.7	5.4
Lys + His + X	14.3	18.5
Phe	0.0	Trace
Pro	4.7	23.7
Ser	2.6	6.3
Thr	4.4	15.3
Tyr	0.0	Trace
Val	0.0	6.9
Subtotal	154.2	354.3
Tau	20.5	27.7
Trimethylamine oxide	45.3	75.8
Betaine	25.7	21.0
Undetermined nitrogen	89.3	131.9
Subtotal	180.8	256.6
Grand total	520.0	1094.1
Blood osmolarity	588.0	1117.6

[a] After Bricteux-Grégoire et al.[10]

olism despite the necessities of low substrate concentrations imposed by the isosmotic properties of the cell. Other factors may rank equally with the isosmoticity of the cell. As discussed by Atkinson,[11] the amounts of molecules and ions that can be present in the cytoplasm must be limited if the solvent capacity of water be retained. However, the metabolism has to remain efficient despite the low intracellular concentrations of intermediate substrates. Therefore, one must find the evolutionary answer to this challenge.

The answer to the question of how an efficient metabolism can be maintained in the presence of low concentrations of intermediary substrates is found in what may be called the principle of spatial correlation. This principle explains how permanent biological structures may have originated in evolution from the dynamic spatial order described by Prigogine.

I. THE PRINCIPLE OF SPATIAL CORRELATION

A metabolic process functioning in a bulk reaction–diffusion medium is slow and inefficient. Remember the slow rate of glycolysis that obtains in the cell-free system described by Hans Buchner at the turn of this century. See also the argument of Pollard[12] showing that formidable concentrations of intermediates would be needed to obtain the actual rate of ethanol production, if the glycolytic enzymes were not associated in the yeast cells. Other examples in point may be found in the consideration of some anabolic pathways: when the enzymes are not physically associated the rate of production with respect to that found in cells where they exist as cluster is 30–100 times lower. Thus, by evolving from a bulk reaction–diffusion system to one topographically segregated into a cluster important kinetic effects are observed. They have been amply discussed by Welch,[13] and their most significant evolutionary implications may be summarized as follows. The first consequence of enzyme organization is the introduction of a vectorial character or metabolic channeling. As a matter of consequence, individual metabolic processes are topographically segregated, competing pathways are separated, low intracellular but high local concentrations of reaction substrates are achieved, a drastic reduction of transient times is observed, and diffusional interferences are eliminated.

Another important feature of enzyme cluster is the attribute of coordinate regulation that may take various forms among which is a regulation by an effector acting at a single site on the cluster but affecting more than one enzyme. Finally, enzyme organization may be shown to flatten the free-energy profile of the reactions of metabolic sequence. The smoothing effect may result from a lowering of energy barriers and/or a raising of the energy valley.

One may try to single out the main pressure channeling the evolution in that direction. Beside the well-known arguments developed by Prigogine and regarding the structuration via energy flow, it seems to me that the necessity of keeping the solvent capacity of water and low concentrations of inorganic ions (due perhaps to their chaotropic effect) together with the prerequisite of cellular isosmoticity are very likely candidates.

II. SUMMARY

Taking into consideration the main pressures channeling the evolution toward metabolic efficiency, that is, increasing free energy dissipation and energetic efficiency, one may single out

1. Formation of cells, that is, isolation of a volume of reactants.
2. Low inorganic ions level (stability of macromolecules, chaotropic agents, protein synthesis, etc.).
3. Preservation of the solvant capacity of water.

It is obvious that the answer is the principle of spatial correlation, that is, the compartmentalization of a metabolism via the formation of oligomeric enzymes, enzyme clusters, and physical association of enzymes in metabolic sequences anchored on membranes.

In contemporary cells, these various aspects are tightly bound, narrowly linked to each other. Take for instance the importance of the so-called sodium pump and of the amino acid metabolism in maintaining the cell volume constant but also in keeping the ionic environment of the cell constant.

As to the mechanism explaining the evolutionary trend toward structuralization and therefore increased free energy dissipation and energetic efficiency, it still remains to be shown how the dynamic spatial order, so well described by Prigogine and so far the only principle that explains the biological order, became frozen in permanent biological structures.

References

1. M. Florkin and E. Schoffeniels, *Molecular Approaches to Ecology,* Academic Press, New York, 1969.
2. E. Schoffeniels, *Biochimie Comparée,* Masson, Paris, 1983.
3. E. Schoffeniels, *Comparative Biochemistry,* Pergamon Press, Oxford, 1984.
4. F. Jacob, *La Logique du Vivant,* Gallimard, Paris, 1970.
5. J. Monod, *Le Hasard et al Nécessité,* Seuil, Paris, 1970.
6. E. Schoffeniels, *Anti-chance,* Pergamon Press, Oxford, 1976.
7. I. Prigogine and R. Lefever, *Adv. Chem. Phys.,* **29,** 1–28 (1975).
8. E. Schoffeniels, In *Living Systems as Energy Converters,* R. Buvet, M. J. Allen, and J. P. Massué, eds., North-Holland, Amsterdam, 1977, pp. 261–270.
9. E. Schoffeniels, *Biochem. Soc. Symp.,* **41,** 179–204 (1976).
10. S. Bricteux-Grégoire, G. Duchâteau-Bosson, C. Jeuniaux, and M. Florkin, *Arch. Int. Physiol. Biochim.,* **70,** 273–286 (1962).
11. D. E. Atkinson, *Curr. Top. Cell Reg.,* **1,** 29–43 (1969).
12. E. Pollard, *J. Theor. Biol.,* **4,** 98–112 (1963).
13. G. R. Welch, *Prog. Biophys. Mol. Biol.,* **32,** 103–191 (1977).

BIFURCATIONS AND SYMMETRY BREAKING IN FAR-FROM-EQUILIBRIUM SYSTEMS: TOWARD A DYNAMICS OF COMPLEXITY

G. NICOLIS

*Faculté des Sciences, Université Libre de Bruxelles,
Bruxelles, Belgium*

I. INTRODUCTION

As explained in the discussion by I. Prigogine, bifurcations under far-from-equilibrium conditions constitute the natural mechanism of evolution and of acquisition of complexity of large classes of systems. Indeed, bifurcation brings about such qualitative changes as the breaking of spatial or temporal symmetries; the emergence of new functions and new morphologies; the ability to choose between different pathways, depending on the initial and external conditions; and an enhanced sensitivity in perceiving the environment and responding flexibly to a change. All these features are characteristic of the complex systems that we observe in nature, of which biological organisms constitute the most striking example.

Here, I would like to elaborate further on the theme of bifurcation. Section II describes the present state of bifurcation analysis of nonequilibrium systems and gives some of my personal perspectives on what I consider to be some promising lines of development. Section III reviews a number of physical, chemical, or biological problems which can be modeled by means of this theory and which provide us with illustrations of chemical evolution, the subject of the present volume. A representative case, the origin and selection of chirality, is analyzed in Section IV. Some conclusions regarding the dynamics of self-organizing systems are presented in Section V.

177

II. BIFURCATION ANALYSIS OF COMPLEX SYSTEMS: WHERE DO WE STAND?

A. The Primary Bifurcation Equations

Primary bifurcation, the first transition from the reference state on the thermodynamic branch, was defined and discussed in the paper by I. Prigogine. This phenomenon is nowadays well understood. Let us outline briefly its theoretical formulation for the reaction–diffusion equations[1,2]

$$\frac{\partial \mathbf{X}}{\partial t} = \mathbf{v}(\mathbf{X}, \{\lambda_i\}) - \text{div } \mathbf{J}_x \tag{1}$$

where $\mathbf{X} = \{X_1, \ldots, X_n\}$ and the flux $\mathbf{J}_x$ is given by Fick's law:

$$\mathbf{J}_x = -\mathbf{D} \cdot \nabla X \tag{2}$$

Let $\mathbf{X}_s$ represent the uniform steady-state solution on the thermodynamic branch. Such a state always exists if $\mathbf{v}$ and $\mathbf{D}$ in equations (1) and (2) do not depend explicitly on space and time and if the system is subjected to symmetric boundary conditions. We introduce the deviation $\mathbf{x}$ from $\mathbf{X}_s$,

$$\mathbf{X}(r, t) = \mathbf{X}_s + \mathbf{x}(\mathbf{r}, t) \tag{3}$$

and write equation (1) in the form

$$\frac{\partial \mathbf{x}}{\partial t} = \mathbf{L}(\lambda) \cdot \mathbf{x} + \mathbf{h}(\mathbf{x}, \lambda) \tag{4}$$

Here $\mathbf{L}$ is the linearized part of the operator appearing in the right-hand side of equation (1), and $h = O(|x|^2)$ stands for the contribution of the nonlinear terms. Suppose that we control a single parameter λ, and let λ_c be a bifurcation point. We place ourselves in the vicinity of this point, a fact that we express as follows:

$$\mathbf{x} = \epsilon \mathbf{x}_1 + \epsilon^2 \mathbf{x}_2 + \cdots$$
$$\lambda - \lambda_c = \epsilon \gamma_1 + \epsilon^2 \gamma_2 + \cdots \tag{5}$$

where ϵ is an appropriate smallness parameter. The dominant part of equation (4) reduces then to the linear equation

$$\frac{\partial \mathbf{x}_1}{\partial t} = \mathbf{L}(\lambda_c) \cdot \mathbf{x}_1 \tag{6}$$

which is equivalent to an eigenvalue problem:

$$x_1 = ae^{\omega_m t}\phi_m(\mathbf{r}) \tag{7}$$

$$\mathbf{L}(\lambda_c)\cdot\phi_m = \omega_m\phi_m$$

The amplitude a, which remains undetermined at this stage is fixed by the solvability conditions imposed on the equations for the higher order terms of the perturbation expansion (5). We arrive in this way at the following set of *bifurcation equations* for the normalized amplitude $\beta = \epsilon a$.

For a bifurcation of steady-state solutions:

$$-(\lambda - \lambda_c)P_1\beta + P_2\beta^2 = 0 \qquad (\gamma_1 \neq 0)$$

or $\hspace{11cm}$ (8)

$$-(\lambda - \lambda_c)P_1\beta + P_3\beta^3 = 0 \qquad (\gamma_1 = 0, \gamma_2 \neq 0)$$

where the P_i's depend on the values of the parameters appearing in the original equations.

For a bifurcation of time-periodic solutions (Hopf bifurcation):

$$[i(\Omega - \Omega_c) - (\lambda - \lambda_c)P_1]\beta + P_3|\beta|^2\beta = 0 \tag{9}$$

where Ω is the frequency of the oscillation and Ω_c its value at the bifurcation point λ_c.

The (generally complex) quantity β plays a role analogous to the *order parameter* familiar from phase transitions. The fact that only one such parameter survives in the final equations illustrates the enormous reduction of degrees of freedom associated with the first bifurcation. Note also the similarity between equations (8) and (9) and the *normal forms* at which one arrives in the qualitative theory of differential equations in the vicinity of resonance points.[3]

B. More Complex Local Bifurcations

As we see from the above equations, the first bifurcation from the uniform steady-state solution is dominated by a single temporal frequency [related to the eigenvalue ω_m in equation (7)] and a single spatial mode [related to the characteristic wavelength of $\phi_m(\mathbf{r})$ in equation (7)]. Thus, although it gives already rise to symmetry breaking and to a beginning of complex behavior, it still enjoys very marked regularity properties. The question arises, therefore, whether one can extend bifurcation theory to describe a whole sequence of transitions generating more complex solutions suitable for representing the dynamics of systems of physical, chemical, or biological interest. As it turns out this extension takes very different forms, according to whether bifurcation remains a *local phenomenon* or becomes "out of control" so to speak and acquires a *global character*. The local aspects of complex bifurcations will be discussed first.

The basic idea is that in order that bifurcation remains a local event, it is necessary that the successive bifurcation points associated with the various transitions be close to each other. To achieve this one introduces additional control parameters μ, ρ, and so on, in such a way that for a suitable choice a *degenerate eigenvalue* of the linearized operator $\mathbf{L}$ is reached.[4] By removing slightly the parameters from this critical value $(\bar{\lambda}, \bar{\mu}, \ldots)$ one obtains bifurcating branches that remain in the vicinity of the reference uniform steady-state solution and can therefore be computed perturbatively.

The simplest new phenomenon induced by this mechanism is a *secondary bifurcation* from the first primary branch, arising from the interaction between the latter and another nearby primary branch. It leads to the loss of stability of the first primary branch or to the stabilization of one of the subsequent primary branches, as illustrated in Fig. 1. The analysis of this branching follows similar lines as in Section I.A, except that one has now two control parameters λ and μ, which are both expanded [as in equation (5)] about the degeneracy point $(\bar{\lambda}, \bar{\mu})$ corresponding to a *double eigenvalue* of the linearized operator $\mathbf{L}$. Because of this double degeneracy, the first equation (7) is replaced by

$$\mathbf{x}_1 = a_1 \boldsymbol{\phi}_{m_1} + a_2 \boldsymbol{\phi}_{m_2} \tag{10}$$

The normalized amplitudes $\beta_i = \epsilon a_i$ $(i = 1, 2)$ obey now a set of two coupled bifurcation equations. As an example, in a one-dimensional re-

Solution

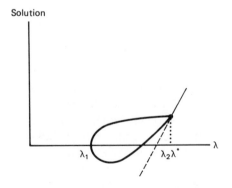

Fig. 1.   Illustration of a secondary bifurcation. λ_1, λ_2, primary bifurcation branches; λ^*, secondary bifurcation point, at which the upper branch bifurcating at λ_2 becomes stabilized.

action–diffusion system subject to zero flux or to periodic boundary conditions and giving rise to a steady-state, the interaction between two modes of wave number one and two, respectively, is described by[5]

$$-[(\lambda - \bar{\lambda})P_1 + (\mu - \bar{\mu})P_1']\beta_1 + P_2\beta_1\beta_2 = 0$$

$$-[(\lambda - \bar{\lambda})Q_1 + (\mu - \bar{\mu})Q_1']\beta_1 + Q_2\beta_1^2 + Q_3\beta_2^3 = 0$$

(11)

All transitions associated with secondary bifurcations have now been completely classified. Moreover, several examples of tertiary or even quaternary branchings are known both for steady-state and for time-periodic solutions (see, e.g., Iooss[6] and Erneux and Reiss[7] for some recent results).

C. Singular Perturbation of Bifurcations. Imperfection theory

In many applications involving nonequilibrium instabilities and dissipative structures, the sharp transitions corresponding to bifurcation rarely occur. Small impurities, imperfections, or external fields tend to distort these transitions. Many experiments, particularly in fluid dynamics, illustrate this fact and demonstrate, in addition, that small imperfections may have large or even qualitative effects. This very general phenomenon is at the basis of the enhanced sensitivity of systems operating near a bifurcation point discussed in the chapter by I. Prigogine.

The most powerful tool for analyzing the influence of imperfections is *singular perturbation*.[8] Consider the general nonlinear problem

$$\Phi(\mathbf{x}, \lambda, g) = 0 \tag{12}$$

in which the parameter g characterizes the magnitude of the imperfection. We assume that, as $g \to 0$, equation (12) reduces to the bifurcation problem considered earlier [see equation (4)]. We call $\mathbf{x}_0(\lambda)$ the solutions of this problem, $\mathbf{x}(\lambda, g)$ their deformation by the imperfection. A natural procedure would be to expand $\mathbf{x}(\lambda, g)$ in power of g:

$$\mathbf{x}(\lambda, g) = \sum_{j=0}^{\infty} \mathbf{x}_j(\lambda)g^j \tag{13}$$

Substituting into equation (12), we thus obtain a set of *linear* equations for $\mathbf{x}_j$ of the form

$$\Phi^0(\lambda)\cdot\mathbf{x}_1(\lambda) = 0$$

$$\Phi^0(\lambda)\cdot\mathbf{x}_j(\lambda) = \mathbf{g}_j(\lambda, \{\mathbf{x}_{j-k}\}) \qquad j \geq 2, 1 \leq k \leq j - 1$$

(14)

where Φ^0 is the linearized operator of the unperturbed problem (with $g = 0$) familiar from the first part of the present section.

The first of these equations admits a solution of the type given in equation (7). In order to specify the amplitude of this solution, which remains undetermined in this stage, one has to solve equations (14) for $j \geq 2$. As the operator $\Phi^0(\lambda)$ admits a nontrivial null solution, one has to verify that these latter equations satisfy suitable solvability conditions. If this is the case for all j, then the imperfection constitutes a smooth perturbation whose effects can be handled by the expansion (13). In particular, bifurcation will subsist and the bifurcation points λ_c will be identical to those determined in the absence of the imperfection.

Now, in many instances it may happen that the solvability conditions are violated for at least one j. The simple expansion given in equation (13), frequently referred to as the *outer expansion*, then fails and must be substituted by an *inner expansion*, in which a nonanalytic dependence on g is allowed. This is achieved by the development

$$\lambda - \lambda_c = \sum_j \gamma_j \epsilon^j$$

$$g = \sum_j g_j \epsilon^j \tag{15}$$

$$\mathbf{x} = \sum_j \mathbf{x}_j \epsilon^j$$

Substituting into equation (12) one then obtains a system of linear equations for $\mathbf{x}_j$ which incorporate the effect of the imperfection already at the dominant order. By working out these relations one finally arrives at the bifurcation equations for the amplitude of the dominant part of the solution, $\epsilon \mathbf{x}_1$.

As an example, suppose that the imperfection consists of taking into account the effect of the gravitational field in the reaction–diffusion equations (1) and (2). In an ideal mixture at mechanical equilibrium the flux $\mathbf{J}_k$ of constituent k becomes[9]

$$\mathbf{J}_k = \mathbf{J}_k(\text{diffusion}) + \mathbf{J}_k(\text{drift}) \tag{16}$$

$$= -D_k \left(\nabla x_k - \frac{1}{k_B T} (1 - v_k \rho) \mathbf{g} x_k \right)$$

Here D_k is Fick's diffusion coefficient, k_B Boltzmann's constant, T the temperature, ρ the total mass density, v_k the partial specific volume, $\mathbf{g}$ the acceleration of gravity. By working out the imperfection theory for

J_k(drift), considered as a perturbation of the Fickian diffusion term, we arrive at the following relation in the case of zero flux boundary conditions[10]:

$$P_0 g - (\lambda - \lambda_c) P_1 \beta + P_3 \beta^3 = 0 \tag{17}$$

This equation is to be compared with the second equation (8): In the absence of the field, $g = 0$, the two equations become identical and predict a symmetric bifurcation from the trivial state $\beta = 0$:

$$\beta = \pm \left(\frac{P_1(\lambda - \lambda_c)}{P_3} \right)^{1/2}, g = 0 \tag{18}$$

But when the field is nonzero the trivial solution is not allowed. Instead, there is always one real nontrivial solution for all values of the bifurcation parameter λ and a pair of other real solutions which exist only for values of λ larger than a certain value $\overline{\lambda}_c$. However, there exists no bifurcation of new solutions from a given branch. This situation is described in Fig. 10 of the paper by I. Prigogine. It provides the basis for understanding the high sensitivity of the system in the vicinity of λ_c and the *pattern selection* introduced by the gravitational field. We come back to this problem in Section IV.

D. Global Bifurcations

In addition to the results of a local nature presented so far, there are some results that describe qualitatively how bifurcating solution branches behave in the large. Generally speaking, it is impossible to study this behavior exhaustively, because the problems are highly nonlinear and therefore intractable. This difficulty is especially apparent when bifurcation is studied as a function of a single parameter. If however two or more control parameters are varied, some of the bifurcations that are global from the standpoint of a single parameter can be formulated in a local way.[3]

A very important example of global bifurcations is the succession of period-doubling transitions, leading from a steady-state solution to time-periodic solutions of increasing period and finally to nonperiodic behavior. This phenomenon has been studied extensively for iterative equations of the form[11,12]

$$x_{n+1} = 1 - \lambda x_n^2 \tag{19}$$

Specifically, it has been shown that the successive bifurcation points λ_k at which the solution switches from period 2^{k-1} to period 2^k have an accumulation point λ_∞ and obey the asymptotic law

$$| \lambda_k - \lambda_\infty | \sim \text{const} \times \delta^{-k} \qquad k \to \infty \tag{20}$$

where δ is universal. This important discovery due to Feigenbaum[11] provides a tempting interpretation of a great number of experimental data from fluid dynamics and chemistry. For a reason that is not yet understood, it would therefore appear that evolution equations of the form of equation (1) give rise, in a certain range of parameters, to a qualitative behavior similar to that corresponding to the dynamical system of equation (19). For a further discussion on this problem, refer to the paper by Jack S. Turner.

Let us now come to bifurcations which can be formulated in a local way. An interesting class are the *codimension two bifurcations*, whereby two control parameters are varied. Consider, for instance, the most general form, also known as *normal form*, of equations involving two variables near a doubly degenerate critical eigenvalue of the linear stability operator[3]:

$$\frac{dX_1}{dt} = X_2 \tag{21}$$

$$\frac{dX_2}{dt} = \lambda_1 + \lambda_2 X_1 + X_1^2 - X_1 X_2$$

Double degeneracy occurs for $\lambda_1 = \lambda_2 = 0$. When λ_1 and λ_2 are different from zero but remain small, the following change of parameters, variables, and time scale:

$$\lambda_1 = \gamma_1 \mu^2$$

$$\lambda_2 = \gamma_2 \mu \tag{22}$$

$$X_1 = \xi_1 \mu, \qquad X_2 = \xi_2 \mu^{3/2}$$

$$\tau = t\mu^{1/2}$$

transforms the system into

$$\frac{d\xi_1}{dt} = \xi_2 \tag{23}$$

$$\frac{d\xi_2}{dt} = \gamma_1 + \gamma_2 \xi_1 + \xi_1^2 - \mu^{1/2} \xi_1 \xi_2$$

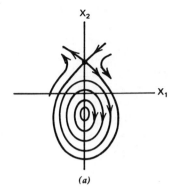

(a)

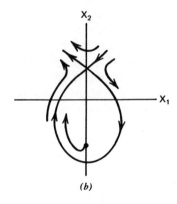

(b)

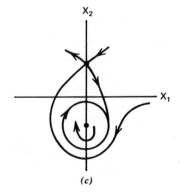

(c)

Fig. 2. Global bifurcation for equations (21)–(23). (a) Hamiltonian reference system. (b) Stable separatrix loop arising from the effect of the "dissipative" perturbation $\mu^{1/2}\xi_1$, ξ_2. (c) Asymptotically stable limit cycle bifurcating from the separatrix loop of (b).

185

This latter system can be viewed as a perturbation of an exactly soluble Hamiltonian system obtained from (23) when μ is set equal to zero.[13,14] The phase portrait of this latter system is depicted in Fig. 2a. Depending on the values of γ_1 and γ_2, one can show that the "dissipative" perturbation $\mu^{1/2}\xi_1\xi_2$ can give rise to a stable separatrix loop surrounding an unstable singular point (Fig. 2b) or to an asymptotically stable limit cycle (Fig. 2c) surrounding an unstable singular point. As γ_2 is varied relative to γ_1, the limit cycle is seen to bifurcate from the separatrix loop, and is thus formed at a finite distance from the singular point. In that respect, it constitutes a global bifurcation,[15] whose period becomes as large as desired if one looks sufficiently near the domain of existence of the separatrix loop.

When equations (21) are driven by a weak external periodic forcing a rich variety of behaviors is observed, the most intriguing of which is the existence of nonperiodic motions displaying a markedly chaotic behavior. Periodic perturbations of Hamiltonian systems have originally been studied by Melnikov[16] and more recently by Holmes[17] and Chow et al.[18] among others. The point we want to make here is that, thanks to the presence of the two control parameters in our starting equations (21), such chaotic motions can be *localized* in the vicinity of the point of double degeneracy $\lambda_1 = \lambda_2 = 0$.[14] In that sense, for certain classes of systems, random nonperiodic behavior becomes a typical phenomenon similar to the bifurcation of steady-state or of time-periodic solutions.

In autonomous systems chaotic behavior is also possible, but can only arise when at least three coupled variables are involved in the dynamics.

III. PHYSICOCHEMICAL EXAMPLES OF BIFURCATION AND COMPLEX BEHAVIOR

The bifurcations summarized in the preceding section endow an initially homogeneous and isotropic system with spatial and temporal degrees of freedom. This phenomenon is referred to as *symmetry breaking* and, as pointed out in Section I, it marks the beginning of self-organization and of complex behavior. That living objects share both of these features is of course hardly anything more than a tautology. Of special significance, therefore, is the fact that simple physical or chemical systems can also show self-organization and complex behavior. The occurrence of these phenomena has been firmly established during the last two decades for such diverse situations as fluids under laboratory or geophysical conditions, chemical reactions, electrical circuits, eutectics, or lasers. All observations corroborate the ideas expressed in the paper by I. Prigogine, namely, that complex behavior arises because of the presence of sufficiently nonlinear feedback interactions and strong nonequilibrium constraints.

Chemical evolution, the subject of the present volume, is the link between the simple behavior of inanimate matter and the complex behavior of biosystems. As discussed in the paper by Stanley L. Miller, the essential preoccupation in this field until now was the prebiotic synthesis of organic molecules and polymers. Considerable progress has been achieved, and new exciting directions such as the investigation of template reactions are under way. In addition to these very important aspects however, we believe that such questions as the possibility that populations of biopolymers may form coherent supermolecular organizations such as, say, a bioenergetic pathway, are equally crucial in understanding chemical evolution. The discovery of bifurcations under nonequilibrium conditions in simple physicochemical systems gives us the tools for searching the beginning of an answer to such questions. In the long run, it should therefore shed new light in the field of chemical evolution.

Let us now give examples of some characteristic forms of supermolecular organization arising through bifurcation in model or in laboratory systems.

A. Sustained Oscillations

In chemistry, this phenomenon arises typically through a primary bifurcation and is well described by equation (9). It has been observed for some time in a variety of systems such as the Belousov–Zhabotinki reaction. New families of chemical oscillators displaying this behavior have also been discovered recently.[19] In biology, oscillations arise through regulatory mechanisms at the enzymatic or the genetic level. They are discussed in the chapter by R. Thomas. In simple fluid dynamic experiments, on the other hand, sustained oscillations correspond to higher order bifurcations. In either case because of the asymptotic stability and the uniqueness of the regime of limit cycle,* one may consider this type of bifurcation as a model of a *time symmetry breaking* establishing an innate rhythmic behavior within the system. In a sense, thanks to the nonequilibrium constraints, the system has been able to "capture" the irreversibility of the evolution equations [cf. equations (1) or (4)] and transform it into an intrinsic, sharply reproducible property.

B. Polarity and Spatial Patterns

The onset of spatial structure in a hitherto homogeneous medium is a central problem in many fields. In embryonic development space struc-

---

* Typically, uniqueness is guaranteed in a certain finite neighborhood of phase space surrounding the limit cycle.

tures provide the necessary "positional" information for cell differentiation, as discussed in the paper by Stuart A. Kauffman. In hydrodynamics
they frequently accompany the onset of convection like in the well-known
Bénard instability. In solid state (metallurgy, crystal growth, snowflakes)
they give rise to intriguing interfacial forms of which dendrites are a striking example.[20] The mechanism of appearance of these patterns is related
to bifurcation, but just which type of bifurcation is involved depends on
the particular pattern considered. Thus, simple quasi one-dimensional
structures (see Fig. 3) displaying polarity usually arise through primary
bifurcation. In other cases involving two- or three-dimensional structures
like in the Bénard problem, higher-order bifurcations are involved.
Higher-order bifurcations may also be responsible for the coexistence of
several patterns of different symmetry such as the polar patterns shown
in Fig. 3 and patterns displaying symmetry around an axis passing through
the middle point of the line $r = L/2$. This interesting possibility is well
described by equations (11), and its implications in developmental biology
have been analyzed by Erneux and Hiernaux.[5]

An important property of the bifurcations leading to pattern formation
is that, for any nonuniform solution $x(\mathbf{r}, t)$, there exists a solution $x(\mathbf{r}', t)$
in which $\mathbf{r}'$ is obtained from $\mathbf{r}$ by the action of a symmetry group $\mathbf{G}$:

$$\mathbf{r}' = \mathbf{Gr} \tag{24}$$

and whose stability properties are identical to those of $x(\mathbf{r}, t)$. The group
$\mathbf{G}$ is the symmetry group of the evolution equations (1) and (2) and is
essentially determined by the geometry of the spatial domain and the
invariance properties of the Laplace operator. In the example of Fig. 3,
$\mathbf{G}$ transforms the solution displaying a polarity axis from, for example,
left to right, into its mirror symmetric image.

The existence of the above-mentioned property confers to the spatially

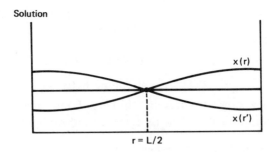

Fig. 3.   Mirror-image polar patterns arising through a primary bifurcation.

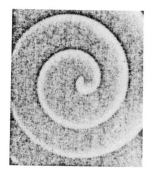

Fig. 4. Spiral wave fronts arising in the Belousov–Zhabotinski reagent (from ref. 31).

asymmetric solutions arising through bifurcation a special status: In a particular physical experiment only one such solution will be realized and one will therefore be allowed to speak of *symmetry breaking*. However, in a statistical ensemble of identical replicas of the system symmetry will be restored, as a particular type of solution will be mixed with equal amounts of its mirror symmetric form. We are faced therefore with an acute problem of pattern selection. We have already referred to this problem in Section I.C, in connection with imperfection theory. A deeper discussion is postponed until Section IV.

C. Chirality and Spatiotemporal Patterns

Time-periodic spatial patterns, usually in the form of propagating wave fronts, have been observed in a wide range of systems: morphogenesis and aggregation of the social amoebae *Dictyostelium discoideum*,[21] cardiac muscle during some forms of arrhythmias,[22] and the Belousov–Zhabotinski reaction.[23,24] It has been suggested[25] that propagating waves should be present in a variety of biological developmental processes to provide the temporal organization of the positional information.

A fascinating feature of many of these wavelike solutions is to present rotating *spiral fronts* (cf. Fig. 4). In an oscillating medium the spirals appear by pairs, the two members of which rotate in opposite directions (clockwise and counterclockwise). We thus have a similar situation as in the previous subsection: Each spiral is an asymmetric pattern displaying a preferred *chirality*, but in a population of spirals in a macroscopic system symmetry is restored by the coexistence of equal numbers of forms of opposite handedness. From the point of view of bifurcation theory two possible origins of spirals have been suggested: In one view, spirals arise from the presence of heterogeneities in the system giving rise to a preferred leading center, around which a wave propagates (see, e.g., ref. 26). And in another view, spirals are intrinsic properties of a system arising through a mechanism of higher bifurcations.[27] The latter view is also supported by work on rotating (nonspiral) waves using the Brusselator model chemical reaction.[28,29]

Spirals have also been shown to arise in a nonoscillating medium.[24,30] In such "excitable" systems, Agladze and Krinsky[31] have established the existence and stability of structures of higher order of symmetry such as the multiarmed vortices shown in Fig. 5. This remarkable phenomenon is quite unexpected from the point of view of the theory of modern critical phenomena, where one shows that vortices with more than one arm (referred to as vortices with "topological charge" $N > 1$) are unstable.[32] It shows that, despite a number of appealing analogies, one should be fully aware of the deep differences between bifurcations in far-from-equilibrium systems and the more familiar phase transition phenomena near thermodynamic equilibrium.

D. Chaotic Behavior

According to Section II.D, under certain conditions the bifurcating solutions of a nonlinear system can display nonperiodic behavior in time. Such phenomena had long been known in fluid dynamics, where turbulence is a ubiquitous feature of large scale flows,[33] and in discrete generation systems such as insect populations.[34] More surprising was the recent discovery of similar behavior in chemistry, particularly in the Belousov–Zhabotinski reaction. We refer to the paper by Jack S. Turner for a discussion of this point. A variety of abstract kinetic models showing chaotic behavior has been proposed by Rössler.[35]

In biology, aside from population systems, chaotic behavior has been suggested to arise in certain pathological conditions resulting from instabilities in physiological control systems,[36] or in the information processing by the brain.[37] In the context of metabolic regulation, a three-variable model describing two positive enzymatic regulatory steps cou-

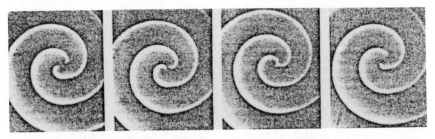

Fig. 5. Successive wave movements of two-armed vortices arising in the Belousov–Zhabotinski reagent (from ref. 31).

pled in series has been analyzed by Decroly and Goldbeter[38] and shown to exhibit chaotic behavior.

An intriguing possibility is the existence of spatially irregular solutions.[39,40] This type of "diffusion induced" chaos certainly warrants further study.

In addition to the transition phenomena mentioned so far in the present section, a variety of even larger scale processes might have operated during chemical evolution, namely, instabilities and bifurcations in the very atmospheric environment within which life emerged. As shown in the paper by Marcel Nicolet, the earth's atmosphere is the theater of a variety of complex chemical and transport phenomena. Moreover, as explained by Stanley L. Miller, the composition of the primordial atmosphere has certainly affected deeply the chemistry in the primitive oceans. Conversely, once life emerged the properties of the atmosphere changed radically, and this must have affected the further course of evolution. We refer to Prather et al.[41] and North et al.[42] for an account of present views on large scale transitions in the earth–atmosphere system.

IV. THE SELECTION OF CHIRAL FORMS

Of all asymmetrical forms of matter, chirality has always exerted a particular fascination. Pasteur's bewilderment over the optical asymmetry of biomolecules, which he considered to be one of the basic aspects of life, is well known. The observation of the morphological asymmetry of adult organisms has introduced into human thinking the notions of "right" and "left," which subsequently turned out to be deeply rooted in a variety of other natural phenomena ranging from elementary particles to astrophysics (see, e.g., Gardner[43]).

The results surveyed in the preceding two sections provide a first clue to the origin of chirality: chiral patterns can emerge spontaneously in an initially uniform and isotropic medium, through a mechanism of bifurcations far from thermodynamic equilibrium (see Figs. 4 and 5). On the other hand, because of the invariance properties of the reaction–diffusion equations (1) in such a medium, chiral solutions will always appear by pairs of opposite handedness. As explained in Sections III.B and III.C this implies that in a macroscopic system symmetry will be restored in the statistical sense. We are left therefore with an open question, namely, the *selection of forms of preferred chirality*, encompassing a macroscopic space region and maintained over a macroscopic time interval.

In this section we sketch a possible answer to this latter question. Within the framework of the deterministic description adopted throughout this paper*, it is clear that what one needs is to go beyond the framework of traditional bifurcation theory, and consider whether suitable perturbations of the bifurcation equations can induce selection. One type of perturbation, associated with gravitational field, was already analyzed in Section II.C, and shown to lead to the selection of polar patterns. Here we want to find the analogue of this mechanism for chiral patterns.

By definition chirality involves a preferred sense of rotation in a three-dimensional space. Therefore, it can only be affected by a modification of the nonscalar fields appearing in the rate equations. For a reaction-diffusion system [equations (1)] these fields are descriptive of a vector irreversible process, namely, the diffusion flux $\mathbf{J}_k$ of constituent k in the medium. According to irreversible thermodynamics, the driving force conjugate to diffusion is

$$\mathbf{X}_{\text{dif}}^k = (\nabla \mu_k)_T - \mathbf{F}_k \tag{25}$$

in which μ_k is the chemical potential of constituent k, and $\mathbf{F}_k$ is the force per unit mass acting on k. In as much as a linear relationship between $\mathbf{J}_k$ and $\mathbf{X}_{\text{dif}}^k$ is accepted, in agreement with the premises of the local formulation of irreversible processes, we must search into ways of modifying $\mathbf{X}_{\text{dif}}^k$, if we want to see an effect on $\mathbf{J}_k$. Thermodynamic analysis[44] shows that such modifications can be achieved by the action of external fields. In a classical nonrelativistic system at rest the only fields that can couple to the conservation equations are the gravitational field $\mathbf{g}$, the electric field $\mathbf{E}$, and the magnetic field $\mathbf{B}$. Of all these couplings, that involving a circularly polarized electric field is the simplest coupling displaying a preferred chirality.[45] The corresponding diffusion flux reads

$$\mathbf{J}_k = -D_k \left\{ \nabla x_k - \frac{z_k}{k_B T} \mathbf{E}(\mathbf{r}, t) x_k \right\} \tag{26}$$

where we follow the notation of (16), and z_k is the charge per unit mass of k. The simplest geometry in which equation (26) becomes effective is a ring subjected to a circularly polarized light propagating in a direction perpendicular to the ring's plane (see Fig. 6). If the angular momentum

---

* Random elements are undoubtedly of considerable importance in the mechanisms of selection. However, the most obvious of such mechanisms, namely, selection through randomly distributed initial conditions, cannot possibly present a coherent character encompassing a large space region and a long time interval. An interesting possibility would be that bifurcation is not a reproducible experiment because, for instance, it needs nucleation involving an exceedingly long time scale. In this case the question of selection would not even arise (see Commentary by Peter Schuster following the paper by Stanley L. Miller).

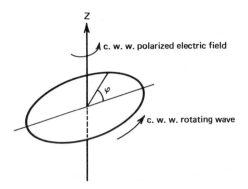

Fig. 6. The coupling described by equation (27).

carried by the wave is along the positive z axis this transverse coupling will result in the following contribution to the diffusion part of equation (1):

$$\left(\frac{\partial x_k}{\partial t}\right)_{\text{dif}} = D_k \left[\frac{1}{R^2}\frac{\partial^2 x_k}{\partial \varphi^2} - \frac{z_k}{k_B T}\frac{E_0}{R}\sin(\omega t - \varphi)\frac{\partial x_k}{\partial \varphi}\right] \quad (27)$$

in which E_0 and ω are, respectively, the amplitude and frequency of the field, R is the radius of the ring, and φ is the polar angle. It was also assumed that the origin of the coordinate system is the center of the ring. Had the incident wave an angular momentum along the $-z$ axis, one would have a term obtained from equation (27), through the substitution $\varphi \to -\varphi$. This is *not* identical to the term displayed in equation (27). As a result, the reaction–diffusion equations for $\{x_k\}$, will no longer be invariant under chirality transformations. By adopting the following compact notation:

$$\mathbf{D} = \left\{\frac{1}{R^2}D_k \delta_{kl}^{kr}\right\}$$

$$\mathbf{M} = \left\{D_k \frac{z_k}{k_B T}\frac{E_0}{R}\delta_{kl}^{kr}\right\} \quad (28)$$

we can write these equations in the form:

$$\frac{\partial \mathbf{x}}{\partial t} = \mathbf{L}(\lambda)\cdot\mathbf{x} + \mathbf{h}(\mathbf{x}, \lambda) - \mathbf{M}\cdot\sin(\omega t - \varphi)\frac{\partial \mathbf{x}}{\partial \varphi} \quad (29)$$

Suppose that the unperturbed system [equation (29) with $\mathbf{M} = 0$] is near a point of bifurcation $\lambda = \lambda_c$ of rotating wave solutions. Following the analysis described in Section II.A, one can show[29] that the dominant part of these solutions near $\lambda = \lambda_c$ is of the form

$$\mathbf{x} = \epsilon \mathbf{x}_1 + \cdots \tag{30}$$

with

$$\epsilon \mathbf{x}_1 = \beta_1 \mathbf{f} e^{i(\Omega t + \varphi)} + \beta_2 \mathbf{f} e^{i(\Omega t - \varphi)} + \text{c.c.}$$

where Ω is the angular frequency and the amplitudes β_1, β_2 obey to *identical* bifurcation equations which have exactly the form of equation (9). Thus, both the "counterclockwise" wave β_2 and the "clockwise" one β_1 bifurcate at exactly the same value λ_c of the control parameter and have identical properties. This had to be expected, in view of our previous discussion on the selection problem.

We want now to see how this state of affairs is affected by the chiral perturbation of our reaction–diffusion equations [term in $\mathbf{M}$ in equation (29)]. To this end we follow the lines of imperfection theory (Section II.C) and expand the variables and parameters in series around $\lambda = \lambda_c$. We also set the frequency Ω of the solution to be identical to the external frequency ω, and assume that ω is close to the linearized intrinsic frequency, Ω_c, in the absence of the field:

$$\mathbf{x} = \epsilon \mathbf{x}_1 + \epsilon^2 \mathbf{x}_2 + \cdots$$

$$\lambda - \lambda_c = \epsilon \gamma_1 + \epsilon^2 \gamma_2 + \cdots \tag{31}$$

$$\omega - \Omega_c = \epsilon \Omega_1 + \epsilon^2 \Omega_2 + \cdots$$

$$\mathbf{M} = \epsilon \mathbf{M}_1 + \epsilon^2 \mathbf{M}_2 + \cdots$$

As stressed in Section II, the coefficients γ_i, $\mathbf{M}_i$ are to be determined from suitable solvability conditions. One finds that to order ϵ^2 the solvability conditions yield $\gamma_1 = 0$. Thus, the amplitudes β_1, β_2 cannot be determined to this order. To order ϵ^3 one obtains a nontrivial result, in the form of two coupled cubic equations for β_1 and β_2. Among the possible solutions of these equations one obtains rotating wave solutions. Setting $\beta_2 = 0$ one finds a clockwise wave:

$$[i(\Omega - \Omega_c) - P_1(\lambda - \lambda_c) + P_1' M^2]\beta_1 + P_3 \beta_1 | \beta_1 |^2 = 0 \tag{32}$$

On the other hand, the equation for the counterclockwise wave β_2, having the chirality of the external field, is

$$[i(\Omega - \Omega_c) - P_1(\lambda - \lambda_c) + P_1''M^2]\beta_2 + M\beta_2(Q_2\beta_2$$
$$+ Q_2'\beta_2^*) + P_3\beta_2 \mid \beta_2 \mid^2 = 0 \quad (33)$$

We see that the presence of the external polarized field destroys the equivalence between the two types of waves.[45] First, the common bifurcation point is split by terms of order of the square of the field amplitude as, in general, $P_1' \neq P_1''$. Second, in the wave having the handedness of the external polarized field there appear quadratic terms in the bifurcation equation which may give rise to subcritical bifurcating branches and to hysteretic behavior. We have therefore succeeded in differentiating between the two kinds of waves, and in a real physical situation this will presumably lead to selection of one of the two forms. Notice that the magnitude of the effect is small, as the naturally occurring polarized fields are very weak. On the other hand as stressed in Section II and in the paper by I. Prigogine, in the vicinity of a bifurcation point a system has a high sensitivity and is thus capable of perceiving very small effects.

To confirm the above conjectures we have performed a numerical simulation of equation (29) on the Brusselator model chemical reaction.[46] The results are shown in Fig. 7. We start with an initial condition corresponding to a clockwise wave. Under the effect of the counterclockwise field this wave is deformed and eventually its sense of rotation is reversed. In other words, the system shows a clear-cut preference for one chirality. As a matter of fact we are witnessing an *entrainment* phenomenon of a new kind, whereby not only the frequency but also the sense of rotation of the system are adjusted to those of the external field. More complex situations, including chaotic behavior, are likely to arise when the resonance condition $\omega = \Omega_c$ is not satisfied, but we do not address ourselves to this problem here.

We have illustrated the selection of chiral patterns using the prototype of reaction-diffusion systems. It is clear however, that our arguments can be extended to include the effect of a variety of other chiral perturbations in more general types of systems. For instance, de Reyff has studied the problem of chirality in a rotating medium[47] and shown that the centrifugal force can also provide a mechanism of selection. Moreover, in the presence of convection a similar role can be assumed by the Coriolis force. These conclusions are interesting, as both types of force are universal perturbations on any earth bound phenomenon like, for instance, the prebiotic synthesis of biopolymers or the spatiotemporal organization of a

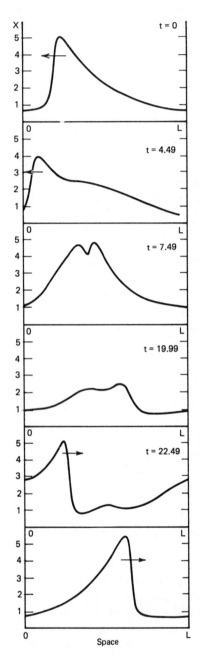

Fig. 7. Numerical simulation of equation (29) for the Brusselator model on a ring. At $t = 0$ a clockwise wave propagates along the ring; at $t = 4.49$ the effect of the counterclockwise external field deforms this wave appreciably; after a while the sense of rotation is reversed and for $t > 22.49$ one obtains a stable wave solution in the counterclockwise direction. The period of both the external field and of the linearized solution of the unperturbed system is 4.28 time units.

196

primitive bioenergetic pathway. Detailed numerical estimations on concrete reaction schemes are needed in order to assess the quantitative importance of these effects.

V. CONCLUDING REMARKS

We have discussed the basis of an *adaptive chemistry,* whereby a system acquires the ability to perceive its environment and interact with it in a sensitive way. Under certain conditions, this interaction allows the system to capture the asymmetry of the external environment and translate it into the properties of a spatial pattern or of a rhythm.

The irreversibility inherent in the equations of evolution of the state variables of a macroscopic system, and the maintenance of a critical distance from equilibrium, are two essential ingredients for this behavior. The former confers the property of asymptotic stability, thanks to which certain modes of behavior can be reached and maintained against perturbations. And the latter allows the system to "reveal" the potentialities hidden in the nonlinearity of its kinetics, by undergoing a series of symmetry breaking transitions across bifurcation points.

Within the medium of broken symmetry generated by this mechanism, physical chemistry is different from the one we are accustomed to. Positional information and internally generated clocks set the stage for the organization of matter at the supermolecular level. In addition, it may be expected that an asymmetric medium acts as "*template,*" enhancing the rate of synthesis of macromolecules of biological interest. True, in the framework of the traditional approach to prebiotic chemistry, extensive experimental evidence on the implications of these conjectures is lacking. Nevertheless, the occurrence of similar phenomena in simple laboratory experiments such as those involving the Belousov–Zhabotinski reagent suggests that symmetry breaking in far-from-equilibrium systems is an important, though still unexplored, mechanism of chemical evolution.

Acknowledgments

I am indebted to Professor I. Prigogine and Drs. M. Herschkowitz-Kaufman and T. Erneux for stimulating discussions. The research reported in this paper is supported, in part, by the U.S. Department of Energy under contract no. DE-AS05-81ER10947.

References

1. G. Nicolis and I. Prigogine, *Self-Organization in Non-equilibrium Systems,* Wiley, New York, 1977.
2. G. Nicolis, in "Systems far from equilibrium," *Lecture Notes in Physics,* Vol. 132, Springer-Verlag, Berlin, 1980.

3. V. Arnold, *Chapitres Supplémentaires de la Théorie des Équations Différentielles Ordinaires,* Mir, Moscow, 1980.

4. J. Mahar and B. Matkowsky, *SIAM J. Appl. Math.,* **32,** 394 (1977).

5. T. Erneux and J. Hiernaux, *J. Math. Biol.,* **9,** 193 (1980).

6. G. Iooss in *Nonlinear Phenomena in Chemical Dynamics,* A. Pacault and C. Vidal, eds., Springer-Verlag, Berlin, 1981.

7. T. Erneux and E. Reiss, to be published.

8. B. Matkowsky and E. Reiss, *SIAM J. Appl. Math.,* **33,** 230 (1977).

9. M. Herschkowitz-Kaufman, G. Nicolis, and A. Nazarea, *Z. Flugwiss. Weltraumforsch.,* **2,** 379 (1978).

10. D. Kondepudi, *Z. Flugwiss. Weltraumforsch.,* **3,** 246 (1979).

11. M. Feigenbaum, *J. Stat. Phys.,* **19,** 25 (1978); **21,** 669 (1979).

12. P. Collet and J. P. Eckmann, *Iterated Maps on the Interval as Dynamical Systems,* Birkaüser, Basel, 1980.

13. J. Keener, *SIAM J. Appl. Math.,* **41,** 127 (1981).

14. C. Baesens and G. Nicolis, *Z. Physik.* **B,** in press.

15. A. Andronov, E. Leontovich, I. Gordon, and A. Maier, *Theory of Bifurcations of Dynamical Systems on a Plane,* Israel Program of Sci. Translations, Jerusalem, 1971.

16. V. Melnikov, *Trans. Moscow Math. Soc.,* **12,** 1 (1963).

17. P. Holmes, *SIAM J. Appl. Math.,* **38,** 65 (1980).

18. S-N. Chow, J. Hale, and J. Mallet-Paret, *J. Diff. Equations,* **37,** 351 (1980).

19. A. Pacault and C. Vidal, eds., *Nonlinear Phenomena in Chemical Dynamics,* Springer Verlag, Berlin, 1981.

20. J. Langer, *Rev. Mod. Phys.,* **52,** 1 (1980).

21. G. Gerisch, *Curr. Top. Dev. Biol.,* **3,** 157 (1968).

22. M. Allesie, F. Bonke, and F. Shopman, *Circ. Res.* **41,** 9 (1977).

23. A. Zaikin and A. Zhabotinski, *Nature (London),* **225,** 535 (1970).

24. A. Winfree, *Science,* **175,** 634 (1972).

25. B. Goodwin and M. Cohen, *J. Theor. Biol.,* **25,** 49 (1969).

26. J. Tyson and P. Fife, *J. Chem. Phys.,* **73,** 2224 (1980).

27. D. Walgraef, G. Dewel, and P. Borckmans, in *Stochastic Nonlinear Systems,* L. Arnold and R. Lefever, eds., Springer-Verlag, Berlin, 1981.

28. T. Erneux and M. Herschkowitz-Kaufman, *J. Chem. Phys.,* **66,** 248 (1977).

29. T. Erneux, *J. Math. Biol.,* **12,** 199 (1981).

30. A. Winfree, in *SIAM-AMS Proceedings,* Vol. 8, AMS, Providence, Rhode Island, 1974.

31. K. Agladze and V. Krinsky, *Nature (London),* **296,** 424 (1982).

32. J. Zeldovich and B. Malomed, *Dokl. Acad. Nauk USSR,* **254,** 92 (1980).

33. H. Swinney and J. Gollub, eds., *Hydrodynamic Instabilities and the Transition to Turbulence,* Springer-Verlag, Berlin, 1981.

34. R. May, *Nature (London),* **261,** 459 (1976).

35. O. Rossler, *Ann. N.Y. Acad. Sci.,* **316,** 376 (1979).

36. L. Glass and M. Mackey, *Ann. N.Y. Acad. Sci.,* **316,** 214 (1979).

37. J. S. Nicolis, *J. Franklin Institute,* in press.

38. O. Decroly and A. Goldbeter, *Proc. Natl. Acad. Sci. USA*, **79**, 6917 (1982).

39. Y. Kuramoto, *Suppl. Progr. Theor. Phys.*, **64**, 346 (1978).

40. L. Howard, *Lect. Appl. Math.*, **17**, 1 (1979); N. Kopell, *Ann. N.Y. Acad. Sci.*, **357**, 397 (1980).

41. M. Prather, M. McElroy, S. Wofsy, and J. Logan, *Geophys. Res. Lett.*, **6**, 163 (1979).

42. G. North, F. Cahalan, and J. Coakley, *Rev. Geophys. Space Phys.*, **19**, 91 (1981).

43. M. Gardner, *The Ambidextrous Universe*, Scribners, New York, 1979.

44. P. Mazur and I. Prigogine, *Mém. Acad. Roy. Belg. Cl. Sci.*, **28**, 1 (1953).

45. G. Nicolis and I. Prigogine, *Proc. Natl. Acad. Sci. USA*, **78**, 659 (1981).

46. G. Nicolis and M. Herschkowitz-Kaufman, to be published.

47. C. de Reyff, *Bull. Cl. Sci. Acad. Roy. Belg.*, **67**, 864 (1981).

COMMENTARY: STOCHASTIC MODELS OF SYNTHESIS OF ASYMMETRIC FORMS

H. L. FRISCH

Department of Chemistry and Physics,
State University of New York,
Albany, New York

In his interesting paper Professor Nicolis raises the question whether models can be envisioned which lead to a spontaneous spatial symmetry breaking in a chemical system, leading, for example, to the production of a polymer of definite chirality. It would be even more interesting if such a model would arise as a result of a measure preserving process that could mimic a Hamiltonian flow. Although we do not have such an example of a chiral process, which imbeds an axial vector into the polymer chain, several years ago we came across a stochastic process that appears to imbed a polar vector into a growing infinite chain.

We are concerned with the distribution of configurational sequences in vinyl polymers[1,2] obtained from polymerizing vinyl monomers such as $CH_2{=}CHR$ or even the symmetrically substituted monomers[3] such as $CHF{=}CHF$. Such a polymerization is always suitably initiated in such a way that the polymer propagates (i.e., is built up linearly) with only one active terminal for further monomer addition. Taking for simplicity the case of $CH_2{=}CHR$, the tetrahedral nature of the carbon bonds implies, that successive R groups in the polymer can either lie on the same side of the plane determined by the connected carbon backbone of the polymer or they can lie on opposite sides of that plane. A notation which is sometimes useful in a printed text is exemplified by (I) . . . bbbpbp . . . in which the orientation of the legs of the letters is intended to portray geometrical relations and only the terminus without the (I) is available for further monomer addition. Thus, one writes

$$(I) \ldots bb \ldots = (I) \ldots pp \ldots = (I) \ldots m \ldots$$

and

$$(I) \ldots bp \ldots = (I) \ldots pb \ldots = (I) \ldots r \ldots$$

where the letters m and r stand for a "meso" dyad sequence and "racemic" dyad sequence of monomers, in conformity with the ubiquitously employed notation of organic chemistry. Thus, a polymer of specific configurational sequence (i.e., tacticity) can be designated as in the following example, (I) . . . mrmmrmrrm . . . , by an infinite sequence composed of the letters m and r.

Various non-Markovian and Markovian models have been proposed to account for observed frequencies of low order n(ads). Restricting ourselves to Markovian propagation models one finds an interesting indeterminacy for sufficiently complex propagation models in that one must consider *reversibility* of a model mechanism.[1] Consider the pure polymer[1]

$$(I) \ . \ . \ . \ mmrmrrmmrmrr \ . \ . \ . \tag{1}$$

which has repeat period six. A hypothetical mechanism generating this polymer (from left to right) is the third-order Markov model defined by the schema of transition probabilities:

$$1 = P(m \mid mmr) = P(r \mid mrm) = P(r \mid rmr) \tag{2}$$
$$= P(m \mid mrr) = P(m \mid rrm) = P(r \mid rmm)$$

Reversing this polymer end for end gives

$$(I) \ . \ . \ . \ rrmrmmrrmrmm \ . \ . \ . \tag{3}$$

which cannot be generated by the above model. In this case, any section consisting of four or more letters allows forms (1) and (3) to be distinguished with certainty. More generally a model may generate polymers that are statistically distinguishable from their reversals. To each such model corresponds a *reversed model,* which may or may not be physically reasonable. A stochastic polymerization mechanism is *completely reversible* if it produces polymers that are statistically indistinguishable from their reversals. A mechanism is *nth-order reversible* if each n(ad) is produced with the same frequency as is the reversed n(ad). End effects are to be ignored in the definition of reversibility.

It can be shown that every theoretical mechanism is automatically nth-order reversible for $n = 1, 2, 3, 4$ since in any polymer all n(ads) of these orders are automatically equifrequent with their reversals. Also, all first- and second-order Markov models are completely reversible. Third (and higher)-order Markov models are not completely reversible in general.

Specifically, a third-order model is reversible iff

$$q(\text{mmr})q(\text{rrm}) = q(\text{mrm})q(\text{rmr}) \qquad (4)$$

where

$$q(\text{xyz}) = \frac{P(\text{m} \mid \text{xyz})}{P(\text{r} \mid \text{xyz})}.$$

To examine the question whether such a stochastic process can be imbedded in a measure preserving flow we proceed as follows: Let us number the possible states of the polymer according to the triad in which the polymer terminates:

1 if . . . mmm, 2 if . . . mmr, 3 if . . . mrm,
4 if . . . mrr, 5 if . . . rrr, 6 if . . . rrm,
7 if . . . rmr, 8 if . . . rmm

Consider the stochastic process [easily seen to be related to that described by equation (2)] whose transition probability matrix $M_{ij} = M(i, j = 1, \ldots, 8)$ is the doubly stochastic matrix

$$M = \begin{pmatrix} 1 - \epsilon & \epsilon & 0 & 0 & 0 & 0 & 0 & 0 \\ 0 & 0 & 1 - \epsilon & \epsilon & 0 & 0 & 0 & 0 \\ 0 & 0 & 0 & 0 & 0 & 0 & 1 - \epsilon & \epsilon \\ 0 & 0 & 0 & 0 & \epsilon & 1 - \epsilon & 0 & 0 \\ 0 & 0 & 0 & 0 & 1 - \epsilon & \epsilon & 0 & 0 \\ 0 & 0 & 0 & 0 & 0 & 0 & \epsilon & 1 - \epsilon \\ 0 & 0 & \epsilon & 1 - \epsilon & 0 & 0 & 0 & 0 \\ \epsilon & 1 - \epsilon & 0 & 0 & 0 & 0 & 0 & 0 \end{pmatrix} \qquad (5)$$

Providing $\epsilon < 1/2$, the equality given by (4) cannot be satisfied. Furthermore there exists an invariant state vector **P** all of whose elements are 1/8 which satisfies

$$\mathbf{MP} = \mathbf{P} \qquad (6)$$

Thus M describes a mixing Markov shift which is not completely reversible. But Friedman and Ornstein[4] have shown that such a mixing Markov shift is isomorphic to a Bernoulli shift, and previously Ornstein[5] has shown that a Bernoulli shift can be imbedded in a measure preserving flow. This answers our basic question.

We remark here that whereas certain polymers, for example, the isotactic polymer (I) . . . mmmmm . . . , crystallize often in a helix form, in a given sample equal amounts of helices of both handedness are present. Hence, we have not generated an example that propagates a polymer of a single helicity. Finally, although one can generate such stochastic games with facility this does not constitute a proof of compatibility with an underlying Hamiltonian dynamics.

Acknowledgments

I am indebted to Drs. G. Nicolis, B. Misra, C. Mallows, F. Bovey, and N. Friedman for many suggestions and clarifying discussions. This work was supported by the NSF grant DMR780593804.

References

1. H. L. Frisch, C. L. Mallows, and F. A. Bovey, *J. Chem. Phys.*, **45**, 1565 (1966).
2. F. A. Bovey, *High Resolution NMR of Macromolecules*, Academic Press, New York, 1972.
3. F. A. Bovey, private communication.
4. N. A. Friedman and D. S. Ornstein, *Adv. Math.*, **5**, 365 (1970).
5. D. S. Ornstein, *Adv. Math.*, **4**, 337 (1970).

COMPLEX PERIODIC AND NONPERIODIC BEHAVIOR IN THE BELOUSOV–ZHABOTINSKI REACTION

JACK S. TURNER

Department of Physics
and
Center for Studies in Statistical Mechanics,
The University of Texas, Austin, Texas

I. INTRODUCTION

During the past decade the Belousov–Zhabotinski (BZ) reaction[1] has been the experimental prototype for self-organization phenomena and dissipative structures in nonequilibrium chemistry.[2] Originally comprising the cerium catalyzed bromination and oxidation of malonic acid by bromate, the BZ system (and others derived from it by substitution for metal ion and/or organic acid) has been found to exhibit a surprising variety of types of nonequilibrium behavior including temporal and spatial oscillations and propagating chemical waves.[1,3–11] In view of the ever-increasing attention that these systems have attracted over the years, it is remarkable that qualitatively new types of behavior continue to be revealed in both experimental[5–11] and theoretical investigations.[6,7,12,13] In this paper I will focus on several complex oscillatory modes that are specific to open-system conditions. In this context experimental observations of spatially homogeneous *bursts of oscillation*[5,9] and of apparent *homogeneous chemical chaos*[8,10,11] will be related to the three main types of homogeneous oscillations that characterize earlier closed-system experiments. Referring to a reduced mechanism of the BZ reaction derived from the Field–Noyes models,[14] I will show how each of the observed phenomena can be understood in terms of this simple model. Finally, I will report some new model predictions for the BZ reaction and discuss the implications for experiment.

In closed system studies of the BZ reaction, three principal modes of homogeneous oscillations have been identified: (1) low-frequency, large amplitude, highly nonlinear (i.e., nonharmonic) relaxation oscillations

(RO, Fig. 3d); (2) higher-frequency, smaller amplitude, quasi-harmonic oscillations (QHO, Fig. 3a); and (3) double-frequency oscillations containing variable numbers of each of the two previous types. By far the most familiar feature of the BZ reaction, the relaxation oscillations of type 1 were explained by Field, Körös, and Noyes in their pioneering study of the detailed BZ reaction mechanism.[15] Much less well known experimentally are the quasiharmonic oscillations of type 2,[4,6] although they are more easily analyzed mathematically. The double frequency mode, first reported by Vavilin et al.,[4] has been studied also by the present author and co-workers,[6] who explained the phenomenon qualitatively on the basis of the Field–Noyes models of the BZ reaction.

In closed-system experiments, the observed oscillations represent transients in the overall monotonic approach to equilibrium, making quantitative study difficult. Under open-system conditions (e.g., in a stirred flow reactor), the kind of control needed to maintain an invariant stationary or oscillating regime, and to vary selectively the principal reactants, can be attained. As expected, oscillations of types 1 and 2 above can be maintained indefinitely in such a reactor. In addition to the two simplest situations, however, two qualitatively new modes of oscillation which were not anticipated have also been observed under these conditions: (4) intermittent bursts of type 1 oscillations separated by periods of quiet (homogeneous) quasi-steady state behavior,[5,9] and (5) mixed frequency oscillations with single occurrences of type (1) separated by a generally variable number of type 2.[8,10] In an experimental study of the ferroin-catalyzed version of the BZ reaction, Schmitz et al.[8a] observed mixed-frequency oscillations ranging from periodic to apparently chaotic as the flow rate was varied. More recently, Hudson et al.[8b,c] and Roux et al.[10] have reported similar observations for the cerium-catalyzed BZ reaction. Nonperiodic oscillations have also been reported by Rössler and Wegmann.[11]

In order to understand such complex oscillatory modes in terms of the underlying nonequilibrium chemistry, it is convenient to begin with a simplified version of the full BZ mechanism[15] which describes the main features of earlier experimental observations of the closed BZ reaction.

II. THE "BUFFERED" BELOUSOV–ZHABOTINSKI REACTION

A. The Oregonator Model of Field and Noyes

Here I consider the Field–Noyes model[15] of the BZ reaction. With reverse reactions included,[14b] it takes the form ("model F")

$$A + Y \leftrightarrows X + P \qquad w_1 = k_1 AY - k_{-1}XP \tag{F1}$$

$$X + Y \leftrightarrows 2P \qquad w_2 = k_2 XY - k_{-2}P^2 \tag{F2}$$

$$A + X \leftrightarrows 2X + Z \qquad w_3 = k_3 AX - k_{-3}X^2 Z \tag{F3}$$

$$2X \leftrightarrows A + P \qquad w_4 = k_4 X^2 - k_{-4}AP \tag{F4}$$

$$Z \leftrightarrows fY \qquad w_5 = k_5 Z - k_{-5}Y \tag{F5}$$

where the rate w_i is given to the right of each chemical step. Here $A = [BrO_3^-]$ and $P = [HOBr]$ are taken in sufficient excess to be effectively constant over reasonable observation times, f is a stoichiometric factor that reflects the composite character of reaction (F5), and the intermediates are $X = [HOBr_2]$, $Y = [Br^-]$, and $Z = 2[Ce^{4+}]$. With $f = 1$ for convenience, the stirred (i.e., spatially homogeneous) BZ system is described by three nonlinear ordinary differential equations

$$\dot{X} = w_1 - w_2 + w_3 - 2w_4 \tag{1a}$$

$$\dot{Y} = -w_1 - w_2 + w_5 \tag{1b}$$

$$\dot{Z} = w_3 - w_5 \tag{1c}$$

With $A = 0.06$ M and the rate constants of Ref. 14b, these equations admit a unique homogeneous steady-state solution (HSS). It is well known that the irreversible "Oregonator"[14a] and its reversible counterpart[14b] exhibit homogeneous limit cycle oscillations for realistic values of rate constants and buffered concentrations. My purpose here is to explore several other features of the reversible model (F) which explain a variety of observed behaviors in closed and open stirred reactors. To that end I begin with the stability properties of the unique HSS, as displayed in the partial "phase" diagram of Fig. 1.

B.   Homogeneous Bifurcations and Nonequilibrium Phase Transitions

When the HSS solution of the chemical rate equations (1a)–(1c) first becomes unstable as the distance from equilibrium is increased (by decreasing P, for example), the simplest oscillatory instability which can occur corresponds mathematically to a Hopf bifurcation. In Fig. 1 the line DCE is defined by such points of bifurcation, which separate regions of stability (I,IV) of the HSS from regions of instability (II,III). Along section a–a', for example, the HSS becomes unstable at point α. Beyond this bifurcation point, nearly sinusoidal bulk oscillations (QHO, Fig. 3a) increase continuously from zero amplitude, eventually becoming nonlin-

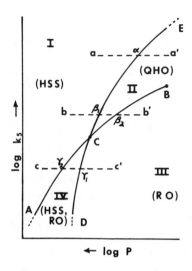

Fig. 1.   Partial (schematic) phase diagram for the reversible Oregonator model of the Belousov–Zhabotinski reaction.

ear (Fig. 3*b* and *c*) as *P* decreases further. The nonequilibrium analogue of a second-order (continuous) phase transition, this situation is indicated schematically in the bifurcation diagram of Fig. 2*a*. Here the amplitude of Br⁻ excursions is plotted against *P* along line *a--a'*. In the stirred BZ reaction, the regime which is finally reached in this way consists of a large-amplitude relaxation oscillation (RO, Fig. 3*d*), which differs qualitatively from both previous types.

In most closed system studies of the BZ reaction, both the onset and the end of oscillatory behavior have been found to be abrupt rather than

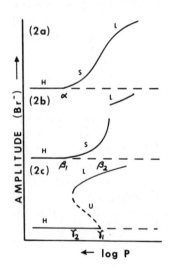

Fig. 2.   Schematic bifurcation diagrams showing the types of transitions along dashed lines *a—a'* (2*a*), *b—b'* (2*b*), and *c—c'* (2*c*) of Fig. 1.

smooth, corresponding to a discontinuous transition between the HSS and the large amplitude limit cycle of the type in Fig. 3d. Such a transition indicates an inverted Hopf bifurcation, analogous to a first-order phase transition. In this case both HSS and RO solutions are stable for a range of values of the external constraints (bifurcation parameter). This is the situation in region IV of Fig. 1, as illustrated by the bifurcation diagram of Fig. 2c, taken along section c––c' of Fig. 1. The possibility of chemical hysteresis in first-order transitions between those two states of different temporal symmetry has been explored already in model studies.[12] This phenomenon is central to understanding certain open-system experiments,[5,9] as will be seen in the next section.

C. Transitions Between Quasiharmonic and Relaxation Oscillations

Closed-system observations of the BZ reaction have revealed that the transition between the small (Fig. 3a–c) and large (Fig. 3d) oscillatory modes may be continuous or discontinuous.[6] In Fig. 1 the transition is continuous for paths passing above point B, for example along section a–a' (cf. Fig. 2a), and discontinuous along segment CB, for example along section b–b' (cf. Fig. 2b). In the language of phase transitions, the QHO ⇆ RO transition is first-order, the "line" of such transitions terminating at the critical point B. In the present macroscopic model [(F1)–(F5)], the width of this first-order transition is effectively zero, precluding even the possibility of hysteresis.

The transition between QHO and RO occurs at a point of numerical discontinuity in the present model (along segment CB of Fig. 1). Hence, one expects that computer error would play a role in determining the stability of oscillatory states near the transition point. The extreme sensitivity of the oscillatory state of the system to parameter values is easily

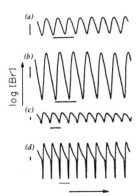

Fig. 3.   Experimental traces of bromide ion concentration in closed system studies of the Belousov–Zhabotinski reaction, showing (a) quasiharmonic (i.e., sinusoidal) oscillations, (b) and (c) increasingly nonlinear oscillations, and (d) relaxation oscillations. The vertical bars at left represent equal concentration ranges.

verified.[21] The transition between the two modes is achieved, in a computer integration of equations (1a)–(1c), by changing a parameter in its thirteenth significant figure. A typical example is shown in Fig. 4, where the time trace of Br^- concentration from a computer integration of the model equations (1a)–(1c), with P = [HOBr] = 0.001 M, is displayed. Initially, with k_5 = 108.6364879365, the system rapidly evolves from any initial condition to a small oscillation (QHO, Fig. 4, $t < 50$ sec). At $t =$ 50 sec the integration is interrupted, then continued with $k_5 =$ 108.6364879364 the only change in the system. Now the stable state is the RO seen in Fig. 4 for $t > 50$ sec. That the RO is then the only possibility is easily verified by starting the integration from any other initial choice of X, Y, and Z values. This great sensitivity provides the first hint that chaotic oscillations may arise from small variations in the "constant" parameter values,[16] or from numerical error in the computer integration of the model equations (1a)–(1c) as suggested in Ref. 19. Indeed, if the relative single-step error exceeds 10^{-10} in the present model, the motion becomes erratic at the QHO–RO transition point, with repeated transitions between the two types of behavior occurring at random.

D. Transitions in the Closed Belousov–Zhabotinski Reaction

If the product P is no longer held (unrealistically) at fixed concentration, then systems at initial points in region III of Fig. 1 will move to the left as P builds up. In this context the three types of terminations observed in typical batch experiments[6,15] are readily understood as trajectories from

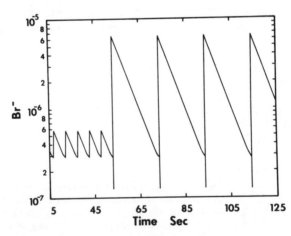

Fig. 4.   Bromide ion trace from a computer integration of model equations (1), showing the response of the system to a small change in a parameter.

region III into region I with passage through region IV, region II, or above point B.[17]

III. COMPLEX OSCILLATIONS IN THE OPEN BELOUSOV–ZHABOTINSKI REACTION

A. A Simple Open-System Model

Following the hint of the preceding paragraph, contact with open-system experiments can now be made quite easily by adding flow terms to a new model with P as a fourth intermediate, and referring to the simpler model phase diagram (Fig. 1) for suggestions of qualitative behavior. In its simplest form, this model treats only the single additional variable P, with concentrations of primary reactions still held constant for convenience. The new model equations are

$$\dot{X} = w_1 - w_2 + w_3 - 2w_4 - \frac{X}{\tau} \tag{2a}$$

$$\dot{Y} = -w_1 - w_2 + w_5 - \frac{Y}{\tau} \tag{2b}$$

$$\dot{Z} = w_3 - w_5 - \frac{Z}{\tau} \tag{2c}$$

$$\dot{P} = w_1 + 2w_2 + w_4 - w_5 - \frac{P}{\tau} \tag{2d}$$

where the flow rate enters as the inverse of the residence time τ (= reactor volume/flow rate), and the chemical fate of P has been inferred from the more detailed mechanism.[18] The key point to recall from studies of the three-variable model [equations (1a)–(1c)] is that the (average) rate of production of P is larger on branch L of Fig. 2a–c than on either of the other stable branches (H or S). If the product P is removed at a rate P/τ such that $\dot{P}_L > P/\tau > \dot{P}_S$ or $\dot{P}_H$, where $\dot{P}$ denotes the contribution of chemical reactions to equation (2d), then the transition regions II and IV become attractors for initial points in region I and III. The result is precisely the kinds of behavior observed in flow-system experiments on the BZ reaction.[5,7–11]

B. Bursts of Oscillations

Consider first the situation along section $c–c'$ in Fig. 1, and begin at the point c. Since $P/\tau > \dot{P}$, P decreases and the system moves into region IV. At the point γ_1, the HSS branch H (Fig. 2c) becomes unstable and an abrupt transition to the RO branch L occurs. Here $\dot{P}_L > P/\tau$, so P begins to build up, and the system moves to the left, executing a number of complete oscillations, and returns past γ_2 to the lower branch H. This hysteresis cycle repeats indefinitely, and the result is the "bursts of oscillation" pattern reported by Sorensen[5] and by DeKepper et al.,[9] and reproduced via model simulation in Fig. 5.

C. Mixed Mode Oscillations

Similarily, along section b–b' in Fig. 1, a system initially at point b will move to the right as P is depleted by flow faster than it is produced. At the bifurcation point β_1 quasiharmonic oscillations commence, increasing in amplitude (and period) as P decreases further along branch S of Fig. 2b. When $P > \beta_2$, branch L is the only stable situation, and one or more relaxation oscillations will occur before P increases beyond β_2 again due to enhanced production rate ($\dot{P} > P/\tau$). In this way the system repeatedly crosses the border between regions I and III, and alternates between two distinct oscillatory regimes. The oscillations which result may be periodic (Fig. 6a and c), or nonperiodic (Fig. 6b) depending on the flow rate. The periodic regime has been found also by Showalter et al.[19] using a slightly more complicated model based also on the reversible Oregonator.

For $\tau < 0.294$ hr and $\tau > 0.305$ hr the oscillations are simple RO and QHO, respectively. If τ is increased from the lower limit in this range, an alternating sequence of periodic and chaotic regimes is revealed. Each periodic regime consists of a single RO and a number of QHO which increases by one from each such periodic regime to the next (Fig. 6a and c). The chaotic states that separate the periodic ones consist of single

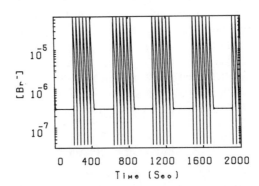

Fig. 5.   Bursts of oscillation observed in a computer integration of the open BZ reaction model equations (2), for $k_5 = 5.0$ and $\tau = 0.926$ hr.

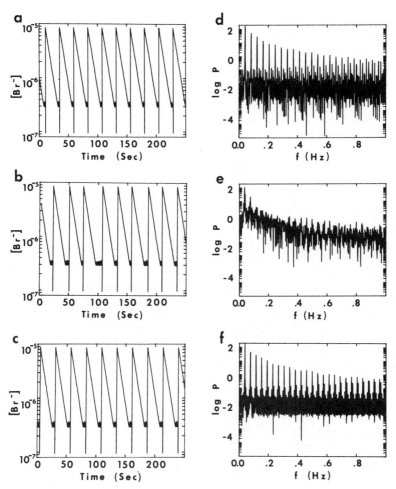

Fig. 6.   Bromide ion traces and power spectra corresponding to three states in the alternating periodic-chaotic sequence, for $k_5 = 90.0$ and τ values (a, d) 0.29535 hr, (b, e) 0.29566 hr, and (c, f) 0.29598 hr.

occurrences of large cycles separated by random numbers of the smaller cycles (Fig. 6b).

In Fig. 6 the Br^- time series for one sequence of alternating states are shown (Fig. 6a–c) together with the corresponding power spectra (Fig. 6d–f). Figure 6a illustrates the first complex periodic state (one RO, one QHO) which appears as τ is increased from 0.294 hr. The second complex periodic state (one RO, two QHO) is shown in Fig. 6c, and the intervening chaotic state in Fig. 6b. Each periodic state is characterized by a power spectrum consisting of a single sharp fundamental frequency component and its harmonics (Fig. 6d and f), while the chaotic states are characterized by a broadband power spectrum (Fig. 6e). The alternating periodic-chaotic sequence has been documented in the model studies up to the 1 RO–8 QHO periodic state. For longer residence times chaotic regimes are found with successively larger numbers of the QHO between single RO, up to a maximum average number of about 100 QHO, before the RO disappear entirely and only QHO remain.

D. Period-Doubling Bifurcations and the Transition to Chaos

For $\tau > 0.3013$ hr, the system remains entirely on the QHO branch S of Fig. 2b. Here a new sequence of periodic and chaotic regimes is predicted by the model studies. This sequence is distinct from the alternating periodic-chaotic sequence found for smaller residence times, and is best understood beginning from the high τ limit of branch S, corresponding to the point β_1 of Fig. 2b. If τ is decreased as discussed in Section III.B, the first oscillatory regime to appear when the system passes from branch H to branch S of Fig. 2b is a simple periodic state, P_1. This state, a "simple" limit cycle like that of Fig. 3a, persists for a range of τ values, Δ_1, in which the amplitude and period increase as the oscillations become more nonlinear (i.e., nonharmonic). It then bifurcates into a second periodic state, called P_2, which has *in each period* two cycles of slightly different amplitude and period. This is the first of a sequence of period-doubling or subharmonic bifurcations in which periodic regimes P_2, P_4, P_8, . . . , P_{2^k} (with k a nonnegative integer) emerge successively. According to Feigenbaum,[20] the intervals Δ_{2^k} decrease geometrically as k increases with an asymptotic ratio $\Delta_{2^k}/\Delta_{2^{(k+1)}}$ (for large k) equal to a universal constant 4.662016 . . . . The bifurcation sequence converges rapidly to an accumulation point corresponding to P_∞, beyond which chaotic regimes appear intermingled with periodic states having odd periods (e.g., P_3, P_5) and period-doubling multiples of those odd periods. In the model studies reported here, the universal constant of Feigenbaum has not been verified directly for computational reasons including, among other things, the phenomenon of critical slowing down near bifurcation points. By way of indirect verification, however, the constant has been used in an operational way to predict the values of τ at which successive P_{2^k} limit cycles should be found, and this procedure has accurately predicted states up to P_{128}. The power spectrum of one such state, P_{64}, corresponding to $\tau = 0.301628$ hr, is shown in Fig. 7. Here three harmonics of the fundamental frequency $f_1 \sim 0.43$ Hz are shown together with the subharmonic components. Further details of the transition sequence predicted by the model are presented elsewhere.[21]

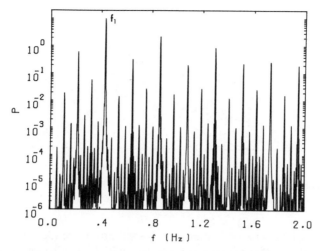

Fig. 7.   Power spectrum for the state P_{64} in the period-doubling sequence, for $k_5 = 90.0$ and $\tau = 0.301628$ hr.

IV.   DISCUSSION

A four-variable model of the open BZ reaction has been analyzed numerically and found to predict several new types of complex periodic and nonperiodic behavior. On the basis of this simple model [equations (2a)–(2d)], each of these complex modes can be understood in terms of repeated passage through a bifurcation region (Figs. 1 and 2) of the higher-level three-variable (reversible Oregonator) model [equations (1a)–(1c)]. By referring to this simpler model, it is possible to explain the sequence of dynamical regimes observed as the residence time is varied in terms of the more familiar simple periodic modes of the experimental BZ system.

The model predictions are summarized in regions $A-E$ of Fig. 8, where the mean amplitude of $[Br^-]$ excursions is plotted schematically against inverse residence time (or flow rate). As this figure indicates, oscillations are found in a domain of τ bounded above and below by regions in which the homogeneous steady state is the stable situation. These are regions A and H, respectively. The other regions delineated by the model studies are B, the region of the period-doubling bifurcation sequence, C, the region of odd-period limit cycles and chaotic states involving small oscillations (QHO), D, the region of mixed-frequency states and of the alternating periodic-chaotic sequence, and E, the region of simple relaxation oscillations. To my knowledge these predictions for the long residence time side of the oscillatory domain have yet to be verified in the laboratory, experiments to date having been concentrated in the region of shorter residence times.

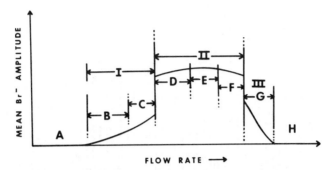

Fig. 8. Schematic composite diagram showing some of the types of behavior observed in experiments on the open BZ reaction and/or predicted by model computations.

The first discovery of a complex transition sequence was made by Hudson et al.[8b,c] The transition sequence which they reported is an alternating periodic-chaotic sequence similar to that discussed in Section III.C, but it has the opposite dependence on residence time. This reflects the fact that the present model studies and Hudson's experiments were carried out on opposite borders of the region in which oscillations are found. The remaining regions of Fig. 8 summarize their observations, with F the region of the alternating periodic-chaotic sequence, and G a second region of smaller amplitude oscillations. No transition sequence (e.g., like that of region B) has been reported in the latter region.

A period-doubling sequence that leads to a regime containing both chaotic and periodic states has been observed quite recently.[22] The general pattern is strikingly similar to that generated by one-dimensional maps with a single extreonum,[20] and occurs within one of the chaotic bands of the alternating periodic-chaotic sequence.

Acknowledgments

This research has been supported by the National Science Foundation through grant CHE79-23627, and by the Aspen Center for Physics during visits in 1979 and 1980.

References

1. B. P. Belousov, *Ref. Radiat. Med.*, **1958**. *Medgiz, Moscow*, 145 (1959); A. M. Zhabotinski, *Dokl. Akad. Nauk. SSSR*, **157**, 392 (1964).

2. G. Nicolis and I. Prigogine, *Self-Organization in Non-Equilibrium Systems*, Wiley-Interscience, New York, 1977.

3. A. M. Zhabotinski, *Oscillatory Processes in Biological and Chemical Systems*, Nauka, Moscow, 1967, p. 149; H. Busse, *J. Phys. Chem.*, **73**, 750 (1969); A. N. Zaikin and A. M. Zhabotinski, *Nature (London)*, **225**, 535 (1970); A. T. Winfree, *Science*, **175**, 634 (1972) and **181**, 937 (1973); D. F. Tatterson and J. L. Hudson, *Chem. Eng. Commun.*, **1**, 3 (1973); M. Marek and E. Svobodova, *Biophys. Chem.*, **3**, 236 (1975).

4. V. A. Vavilin, A. M. Zhabotinski, and A. N. Zaikin, in *Biological and Biochemical Oscillators*, B. Chance, E. K. Pye, A. K. Ghosh, and B. Hess, eds., Academic Press, New York, 1973, p. 71.

5. P. G. Sorensen, *Proc. Faraday Soc. Symp.*, **9**, 88 (1974).

6. J. S. Turner, E. V. Mielczarek, and G. W. Mushrush, *J. Chem. Phys.*, **66**, 2217 (1977).

7. K. R. Graziani, J. L. Hudson, and R. A. Schmitz, *Chem. Eng. J.*, **12**, 9 (1977).

8. (a) R. A. Schmitz, K. R. Graziani, and J. L. Hudson, *J. Chem. Phys.*, **67**, 3040 (1977); (b) J. L. Hudson, M. Hart, and D. Marinko, *J. Chem. Phys.*, **71**, 1601 (1979); (c) J. L. Hudson, D. Marenko, and C. Dove, Discussion meeting, Kinetics of Physicochemical Oscillations, Aachen, September, 1979.

9. P. DeKepper, A. Rossi, and A. Pacault, *C. R. Acad. Sci. Paris*, **283C**, 371 (1976).

10. J.-C. Roux, A. Rossi, S. Bachelart, and C. Vidal, *Phys. Lett. A* **77**, 391 (1980).

11. O. E. Rössler and K. Wegmann, *Nature (London)*, **271**, 89 (1978); K. Wegmann and O. E. Rössler, *Z. Naturforsch.*, **33a**, 1179 (1978).

12. I.-D. Hsu and N. D. Kazarinoff, *J. Math. Anal. Appl.*, **55**, 61 (1976); J. S. Turner, *Phys. Lett. A*, **56**, 155 (1976); J. J. Tyson, *J. Chem. Phys.*, **66**, 905 (1977).

13. J. J. Tyson, *J. Chem. Phys.*, **67**, 4297 (1977).

14. (a) R. J. Field and R. M. Noyes, *J. Chem. Phys.*, **60**, 1877 (1974); (b) R. J. Field, *J. Chem. Phys.*, **63**, 2289 (1975).

15. R. J. Field, E. Körös, and R. M. Noyes, *J. Am. Chem. Soc.*, **94**, 8649 (1972).

16. See, for example, D. Ruelle, *Ann. N.Y. Acad. Sci.*, **316**, 408 (1979).

17. J. S. Turner, E. V. Mielczarek, and J. R. Creighton, to be published.

18. For consistency, the product HOBr must be consumed in the forward reaction of (F5) (as a precursor of bromomalonic acid), although at reasonable levels of HOBr the rate of (F5) does not depend on P explicitly. For further discussion of this point see R. M. Noyes, *Proc. Faraday Soc. Symp.*, **9**, 89 (1974).

19. K. Showalter, R. M. Noyes, and K. Bar-Eli, *J. Chem. Phys.*, **69**, 2514 (1978).

20. M. J. Feigenbaum, *J. Stat. Phys.*, **19**, 25 (1978).

21. J. S. Turner, (a) "Thermodynamics, Dissipative Structures, and Self-Organization in Biology: Some Implications for Biomedical Research," in *Dissipative Structures and Spatiotemporal Organization Studies in Biomedical Research*, G. P. Scott and J. M. McMillin, eds., Iowa State University Press, Ames, 1980, p. 11; (b) "Self-Organization in Nonequilibrium Chemistry and in Biology," in *Self-organization and Dissipative Structures*, W. Schieve and P. Allen eds, University of Texas press, Austin, Texas 1982, p. 40. (c) Discussion Meeting, Kinetics of Physicochemical Oscillations, Aachen, September, 1979.

22. R. H. Simoyi, A. Wolf, and H. L. Swinney, Phys. Rev. Let. **49**, 245 (1982).

BIFURCATIONS IN INSECT MORPHOGENESIS

STUART A. KAUFFMAN

Department of Biochemistry and Biophysics,
School of Medicine,
Philadelphia, Pennsylvania

I. INTRODUCTION

Two of the most fundamental problems in developmental biology are the manner in which cells in different regions of an embryo come to adopt different developmental programs, and the relation between such different programs. The purpose of this chapter is to indicate how bifurcations in reaction-diffusion systems may offer new insights into these problems. The discussion will center on the fruit fly, *Drosophila melanogaster*.

Drosophila is a holometabolous higher Dipteran, possessing egg, larval, pupal, and adult stages.[1] The egg is an ellipsoid about 500 μm long. Following fertilization and egg deposition, the initial zygotic nucleus undergoes 13 nuclear mitoses without division of the egg itself, thereby creating a syncytium. By the ninth division, most nuclei migrate from the yolky core of the egg and arrive nearly synchronously at the cortex. After four further divisions, cell membranes extend down then beneath each nucleus, creating the first true cells which comprise the ellipsoidal mono-layered cellular blastoderm.[2] Shortly thereafter, gastrulation movements begin, with the formation of bilaterally symmetric cephalic and hind gut furrows, which roughly divide the egg longitudinally into thirds, and the ventral midline furrow through which prospective mesodermal tissue invaginates. The tissues that shall form the embryo proper, the germ band, deform the embryo by extending from both polar areas back along the dorsal midline, then the germ band contracts, revealing the initially segmented embryo with 12 longitudinal segments.[3] At 24 hr the embryo hatches as a first instar larva. After three instars, lasting a total of 5 days at 25°C, the third instar larva forms a pupa, the complex processes of metamorphosis ensue, and the adult emerges about 5 days later.

Part of the fascination with *Drosophila* is due to the fact that the entire ectoderm of the adult derives from the terminal metamorphosis of larval

219

organs called imaginal discs (reviewed in refs. 1 and 4). Each disc carries
a tissue and apparently cell heritable determination to form a specific part
of the adult ectoderm: eye, antenna, wing-thorax, first, second, or third
leg, genital structures, etc. (Fig. 1). The operational definition of determination lies in experiments in which a given disc, for example, the wing-
thorax disc, is dissected from a third instar larva and cultured in an adult
abdomen, where the disc cells proliferate, but do not undergo terminal
differentiation to produce adult cuticle. After a 7–14 day period of culture,
the disc implant may be recovered from the host abdomen, cut in two,
and each fragment transplanted to a second host adult, thereby establishing a tissue culture. At any stage, the differentiative capacity of a
subfragment may be assayed by injection into a third instar larval host.
The injected disc fragment undergoes metamorphosis with the host, and
can be recovered as a mass of adult cuticle from the abdomen of the

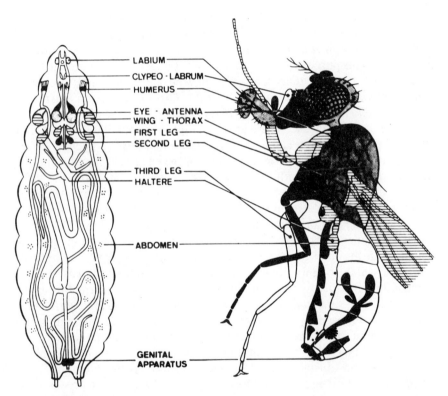

Fig. 1.   Schematic representation of the larval organization and the location of the different
discs. Discs and their corresponding adult derivatives are connected by lines and have the
same hatching or shading.

emerged adult. From the characteristic patterns of hairs, sensillae, and bristles, the implant can be diagnosed as wing, first leg, genital, etc. The major result found by Hadorn and co-workers[5] in such experiments is that a wing disc, or leg disc, subcultured for years, can still metamorphose into adult wing, or leg tissue. Important conclusions can be drawn from these series of experiments. First, disc subfragments have a heritable determination to form a specific adult structure. Second, the heritability is stable over hundreds of cell divisions. Therefore, the molecular mechanisms mediating that heritability could not be merely the initial partitioning to imaginal disc cells of some substance, since it would be diluted out over cell generations. Whatever carries the determined state must be regenerated over cell cycles. Although the mechanisms remain unknown, plausible postulates include integration of transposable genetic elements at diverse loci; chromatin ligands, which bind tightly to genetic control loci, are replicated or replenished once per cell cycle, and are partitioned in orderly ways to daughter cells; and systems of genes and diffusible products which activate and repress one another and possess alternative stable patterns of activity.

The progenitor cells of the different imaginal discs are now known to lie at well-defined positions on the cellular blastoderm.[6-8] Figure 2 shows the blastoderm fate map of *Drosophila*. The features to stress at this stage are that the map is a well-ordered two-dimensional array of pattern elements, bilaterally symmetrical on the closed curved surface of the egg, and that the pattern elements on the fate map are essentially homotopic to the adult. The head maps at the anterior end, followed by the three thoracic fragments, the seven abdominal segments, and the genital segment. Ventral (leg and abdominal sternites) structures map ventrally on the egg to dorsal (thoracic and abdominal tergite) structures. This ordered array of progenitor zones raises the fundamental question of *positional information*.[9] What kinds of processes underlie the establishment of ordered developmental commitments to these heritably different fates? Is there a micromosaic of molecular determinants laid down at specific positions in the oocyte by the mother? Is position in the egg specified by more general "gradients" which she preestablishes? Finally, is position specified "epigenetically" during the early stages of embryogenesis by processes within the early embryo? The answers are not clear,[10-13] but I will sketch in a later section the role dissipative structures may play.

If the existence of an ordered fate map poses the problem of positional information, the striking phenomena of *transdetermination* and *homeosis* unavoidably raise the question of the relation of developmental programs in diverse imaginal discs. Transdetermination was discovered when imaginal disc fragments were cultured. Although each disc normally differ-

STUART A. KAUFFMAN

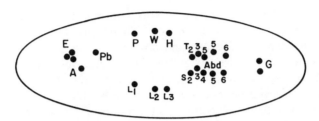

Fig. 2. A, antenna; E, eye; Pb, proboscis; P, prothorax; W, wing and mesothorax; H, halters; L1, L2, L3, first, second, and third legs; Abd, abdominal segments; T2–T6, second to sixth abdominal tergite; S2–S6, second to sixth abdominal sternite; G, genital.

entiates into structures it was initially determined to form, occasionally it forms normal structures from a different disc.[5] A broad body of evidence shows that this alteration in adult derivatives is not due to somatic mutations, and is heritable in the cultured imaginal tissue once established.[1,5] The term *transdetermination* reflects the belief that the alteration is an epigenetic change to a new heritable developmental program. Figure 3 shows the known patterns of transdetermination from each disc. Lengths of arrows reflect probabilities of transdetermination. The following generalizations hold[14]: (1) Most transdetermination steps are reversible, but with asymmetric transition probabilities. (2) Each disc can transdetermine into one or a few, but not all other discs in a single step, thus there are allowed and forbidden one step transitions. (3) Pathways of sequential transdetermination exist, for example, genital to leg to wing to thorax. (4) There is a global ordering towards mesothorax; all transdeterminations that move one step closer to mesothorax are more probable than their inverse.

Transdetermination is an example of metaplasia, the transformation of tissue normally destined for one fate, to a distinct fate. The second major metaplasia in *Drosophila* is due to a class of mutants called homeotic

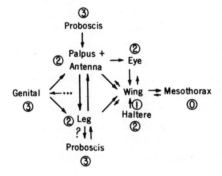

Fig. 3. Patterns of transdetermination from each disc. Length of arrows reflects transdetermination probabilities. Circled numbers, minimum number of transdetermination steps to mesothorax.

mutants (reviewed in Ref. 15). These dominant (or recessive) point, deletion, inversion, or translocation mutants convert one structure to another. Exemplars include the dominant *Nasobemia,* which converts antenna to the mesothoracic leg and less frequently, the eye to wing[16,17]; *tumorous head,* which converts head and antenna tissues to leg, abdominal and genital structures[18]; and the *bithorax* complex which converts parts or all of some thoracic and abdominal segments into one another.[19,20] On the basis of their phenotypes, it is useful to classify homeotic mutants as parallel, divergent, or convergent.[13,21] Parallel homeotics convert a single tissue in a single direction, but may act coordinately on more than one tissue. For example, *Nasobemia* converts antenna to mesothoracic (second) leg and eye to wing (mesothorax). Divergent homeotics, such as *tumorous head,* appear to cause a single tissue to diverge to two distinct fates; antenna to abdomen and leg. Convergent homeotic mutants can transform two different tissues to a common final fate, as seen in *Extrasexcomb,* which converts second and third legs to first legs.[22] As discussed below, these different classes appear to require different hypotheses about the underlying mutant action. A second classification of homeotic mutants asks whether the mutant transforms a tissue to a neighboring domain on the blastoderm fate map, or to a more distant location. *Nasobemia,* for example, largely transforms head and antenna to second thoracic structures, skipping the intervening gnathal and first thoracic structures; similarly, *tumorous head* transforms head to genital and abdominal fates, hence from one end of the fate map to the other. In contrast, *bithorax* alleles largely transform between neighboring areas on the fate map, for example, mesothorax to metathorax.

The importance of transdetermination and homeosis is that these metaplasias presumably reflect "neighboring" relations between the developmental programs in the discs which transform into one another. In some sense then, antenna and leg are program neighbors. Genital, though close to both antenna and leg, is farther from eye, wing, or haltere. It is clear that these "neighboring" relations bear no simple relation to the distance separating different progenitor zones on the fate map, since genital and antenna appear to be developmental neighbors, but derive from opposite poles of the blastoderm.

The observations discussed so far lay out some of the desiderata of any good theory; to try simultaneously to understand how position may be specified in the early embryo, and also to account for the ways in which apparently neighboring developmental programs are not spatial neighbors on the blastoderm.

An important set of clues have come from analysis of the development of the wing-thorax disc. Cells in neighboring domains which have come to adopt different developmental commitments, say to wing or leg, should not thereafter jointly form either a wing or a leg. Further, since determination is cell heritable, it is to be expected that the sets of daughter cells (clones) derived from such domains with different commitments will also not form joint structures. Such domains of cells might be expected to give rise to lines of "clonal restriction," where clones with one commitment on one side would be precluded from crossing to the other side. Such lines, called compartmental boundaries,[23] have recently been found in *Drosophila* development, utilizing the genetic technique of somatic recombination,[24] which allows all the daughter cells (clone) derived from

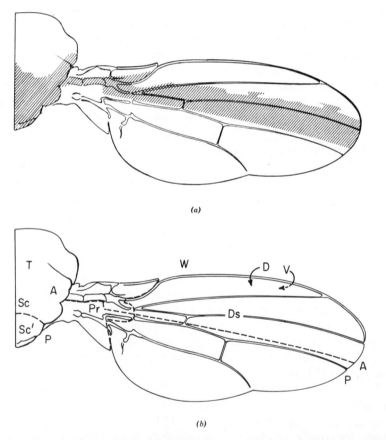

(a)

(b)

Fig. 4. (*a*) A large clone whose posterior margin runs along the anterior–posterior border of the wing and thorax. From Garcia-Bellido et al.[23] (*b*) The five compartmental lines on the thorax and wing. A, anterior; P, posterior; V, ventral; D, dorsal; T, thorax; W, wing; Pr, proximal wing; Ds, distal wing; Sc, scutum; Sc′, scutellum. From Garcia-Bellido[27].

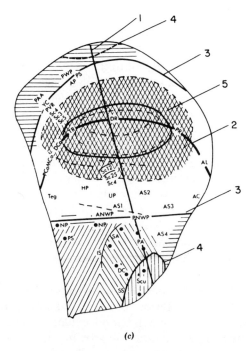

(c)

Fig. 4. (*continued*) (c) Projection of the five compartmental boundaries onto the fate map[26] of the third instar wing disc. The dotted line 4 is the postulated compartmental line in the ventral thorax needed to complete the anterior–posterior, dorsal–ventral (twofold) symmetry.

a randomly affected initial cell to be identified in the adult. The most detailed data for such lines has been amassed in the wing-thorax (hereafter, wing) disc, and I shall focus on this tissue in the present chapter.

At the blastoderm stage, the progenitor zone for the wing disc has about 40 cells.[25] During embryogenesis, this buds inward to form a hollow ellipsoid sac of cells which remains attached to the larval trachea by a narrow stalk. During the three larval instars, the wing disc grows in size and cell number to a final 60,000 cells.[26] In one "hemisphere" of the ellipsoid sac, cells fail to proliferate and stretch to form the thin peripodal membrane that is lost during metamorphosis. The remaining hemisphere of columnar cells forms the disc proper and metamorphoses into the adult wing and hemithorax.

If a cell is genetically marked at about the blastoderm stage, its clonal derivatives are found not to cross a specific line on the adult wing thorax which divides the tissue into anterior and posterior compartments (Fig. 4a). If one examines a number of flies, however, one finds that clones initiated at that stage may cross anywhere else inside either compartment. About mid-first instar, a second clonal restriction occurs and isolates clones to the dorsal or ventral halves of the wing-thorax. At about this time, each clone is further restricted to the wing or the thorax by a third boundary. Somewhat later, a fourth line subdivides the dorsal, and perhaps ventral thorax. In late third instar, a final line isolates the distal from the proximal wing (Fig. 4b). The basic phenomenon, then, is the *sequential* formation of clonal restriction lines which successively subdivide the wing disc into finer subdomains as the disc grows larger in size. The locations of these lines on the adult, however, reveal little of their geometry at the times of their formation. A better approximation is seen by projecting them onto the known fate map of the third instar wing disc (Fig. 4c). The wing disc forms the wing, by folding over the dorsal–ventral line which becomes the adult wing margin, thereby apposing dorsal and ventral thorax areas, and dorsal and ventral wing areas, creating a "bag" that everts back through its peripodal membrane to form the adult wing and thorax.

The striking feature of the compartmental lines on the wing disc is their clear twofold symmetry. Anterior, posterior, ventral, and dorsal quandrants look fundamentally similar. Good evidence suggests that these compartmental lines arise *in situ* in the growing disc, and do not reflect movement of cells relative to one another. Therefore, the question arises of how such symmetrical lines might arise.

II. A DYNAMICAL MODEL

The symmetries and patterns of the lines shown in Fig. 4c are closely analogous to the nodal lines in the eigenfunctions of the Laplacian operator on an ellipse.[28,29] The pleasant question, "Can one hear the shape of a drum?" can be transformed to ask whether one can conceive of an underlying dynamical system having the property that as the size and shape of the wing disc grows, a sequence of different eigenfunctions, with different nodal lines can arise. A wide variety of different underlying mechanisms, having similar spatial operators, and hence similar eigenfunctions, would probably suffice. Figure 5a–f show the nodal lines that would arise on a freely suspended elliptical plate excited to resonant vibration by a gradually higher pitched external sound source.[29] Essentially the same sequence will arise using a reaction-diffusion dissipative structure model which I now discuss briefly.

Perhaps the most general reason to consider reaction-diffusion systems and solutions to the Laplacian is that positional information in *Drosophila* and many organisms seems to be locally averagable. Juxtaposition of tissue edges that are usually not adjacent leads to the intercalary regeneration of pattern elements normally lying between the abutted surfaces.[30-32] The oldest, and still simplest, hypothesis to account for this is to postulate "gradients" of "morphogens" whose concentration specifies position. Juxtaposition of normally nonadjacent edges creates gradient discontinuities followed by a "diffusive-like" smoothing to recreate intervening gradient levels. Therefore, solutions to the Laplacian afford an intelligent first guess to the spatial shapes which such gradients might take.[33] Gradients might be generated by a variety of means. Older hypotheses included special source or sink regions, polarized transport, or other mechanisms. However, recent evidence in *Drosophila* and other systems, described elsewhere[34,35] suggests reaction-diffusion systems as a particularly valuable class of hypotheses, to establish gradients controlling positional information.

Appropriate coupling of reactions and diffusive transport in a planar sheet of cells, considered as a continuum, can readily generate a succession of differently shaped gradients of the same underlying morphogens as the tissue changes in size and shape, or other parameters alter.[21,36] In Fig. 6, we assume specific nonlinear reaction laws, and a simple diagonal diffusion matrix, such that each species' diffusion is independent of the remainder. The dynamical system possesses a spatially homogeneous, temporal steady state. The resulting equations were linearized about this steady state and analyzed in the familiar manner. As it is now well understood (Fig. 6) and discussed in more detail in the Appendix, with appropriate coupling in the linearized reaction terms, and dissimilarity in the diffusion constants of at least two hypothetical morphogens, a dynamical system can be constructed which acts as an amplifier, selecting from the thermal fluctuations around the steady state, a limited range of spatial wavelengths (L1-L2) which will grow in amplitude in time.[21,36-40] If such a chemical system is embeded in a specific geometric domain, and no-flux boundary conditions are imposed (Fig. 6), then only when the size and shape of the domain allow a particular chemical pattern of the amplifiable wavelength to satisfy the boundary conditions will that pattern grow. If the chemical system specified in Fig. 6 were placed in a circular petri dish that gradually enlarged from an initial small radius, the following general results would occur in the *linearized* system of equations: (1) at sufficiently small radii, diffusion overwhelms reaction induced instabilities, and the system remains at the homogeneous state. As the radius of the circular petri dish enlarges beyond a first bifurcation value, given by the first 0 in the derivative of the first Bessel function for wavelength L1, a first chemical pattern arises (Fig. 7a). If the range of amplifiable wavelengths, L1-L2 (Fig. 6), is sufficiently small, then when the petri dish

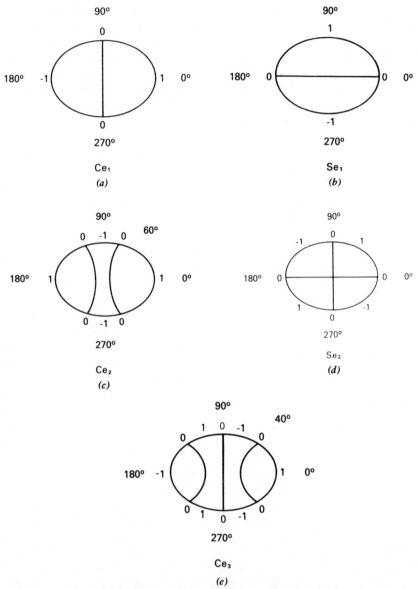

Fig. 5. Nodal lines of successive wave patterns which fit onto an ellipse as it enlarges. The patterns are similar to those on a circle. (a) $Ce_1(\xi, s_{11}) Ce_1(n, s_{11})$ (abbreviated Ce_1) is a slight distortion of $J_1(kr) \cos \phi$ (Fig. 7a). (b) $Se_1(\xi, s_{11})se_1(n, s_{11})$ (i.e., Se_1) is Ce_1 rotated 90°. (c) Ce_2 is analogous to $J_2(kr) \cos 2 \phi$ (Fig. 7b) but on an ellipse the radii split to form pairs of confocal hyperbolas. (d, e) Se_2 and Ce_3 are analogous to $J_2(kr) \sin 2\phi$ and $J_3(kr) \cos 3\phi$, respectively.

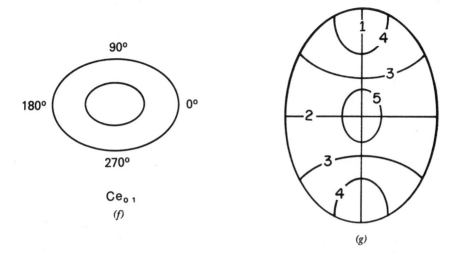

Fig. 5   (*continued*) (*f*) $Ce_0(\xi, s_{01})ce_0(n, s_{01})$, or Ce_0 is analogous to $J_0(kr)$ cos 0ϕ. This mode is a hill shaped pattern with an interior nodal ellipse similar to Fig. 7*c*. See the Appendix for definitions of the symbols. (*g*) Taking account of wing disc growth, we project all five predicted compartmental boundaries onto one ellipse. The observed boundaries are shown in Fig. 4*c*.

increases past a second critical radius, such that longer wavelength, L2, no longer satisfies the boundary conditions, the first chemical pattern will decay toward the homogeneous steady state. As the size of the dish increases beyond a third critical bifurcation value, given by the first 0 in the derivative of the second Bessel function for the smaller wavelength, L1, a second chemical pattern (Fig. 7*b*) will arise, then decay again as the dish continues to enlarge such that the longer wavelength, L2 in the second pattern, does not satisfy the boundary conditions. The third pattern to arise (Fig. 7*c*) corresponds to a radius at which L1 satisfies the second 0 in the derivative of the 0th Bessel function. In these linearized equations, when the radius becomes sufficiently large, two successive chemical patterns will simultaneously satisfy the boundary conditions and both will be amplified by the dynamical system. Thus, their superposition will grow. For smaller circles, the existence of superposition of modes depends on the velocity with which the circle increases in radius relative to the rate of establishment and decay of successive chemical patterns. If the range of allowed wavelengths L1-L2, is sufficiently small, the growing petri dish harbors very nearly a sequence of pure eigenfunction pat-

SUCCESSIVE EIGENFUNCTION MODEL OF COMPARTMENTALIZATION

NONLINEAR REACTION DIFFUSION SYSTEM:

$$\frac{\partial X}{\partial T} = F(x,y) + D_x \nabla^2 x$$

$$\frac{\partial Y}{\partial T} = G(x,y) + D_y \nabla^2 y$$

LINEARIZE ABOUT SPATIALLY HOMOGENEOUS STEADY STATE AND PERFORM STABILITY ANALYSIS

$$\begin{vmatrix} A_{11} - K^2 D_x - \lambda & A_{12} \\ A_{21} & A_{22} - K^2 D_y - \lambda \end{vmatrix} = 0$$

UNDER CONDITIONS:

(i) $A_{11} + A_{22} < 0$

(ii) $A_{11} A_{22} - A_{12} A_{21} > 0$

(iii) $(A_{11} - A_{22})^2 > -4A_{12}A_{21}$

(iv) $D_x A_{22} + D_y A_{11} > 0$

(v) $\left(\sqrt{\frac{D_x}{D_y}} A_{22} - \sqrt{\frac{D_y}{D_x}} A_{11} \right)^2 > -4 A_{12} A_{21}$

THE REACTION-DIFFUSION SYSTEM HAS SPATIAL PATTERNS WITH A "NATURAL" WAVELENGTH L_*

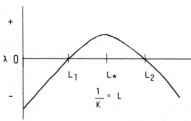

Fig. 6.   Outline of the successive eigenfunction model. Chemical patterns grow in domains when no flux boundary conditions are satisfied and $\lambda > 0$. For more detail, see Appendix.

230

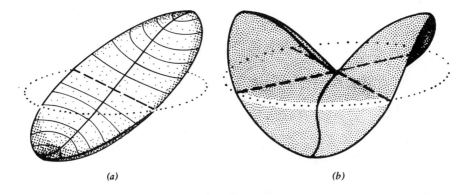

(a) *(b)*

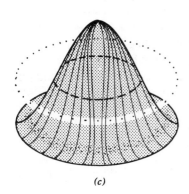

(c)

Fig. 7. (a) Wave pattern generated on a circle with scaled radius $kr = 1.82$. The pattern is the product of a radial part, $J_1(kr)$ [the first-order Bessel function] and an angular part, cos ϕ. The dashed nodal line of zero (i.e., steady state) concentration runs along diameter of the circle from $\phi = 90°$ to $\phi = 270°$. The dotted circle outlines the circular radius. (b) Wave pattern, $J_2(kr)$ cos 2ϕ, generated at a scaled radius of 3.1. The dashed lines are crossed nodal lines on two perpendicular diameters. (c) Pattern generated at a scaled radius of 3.8, where the zero in the derivative of $J_0(kr)$ matches the radial boundary condition. The pattern is $J_0(kr)$ cos 0ϕ, which has no angular variation. The nodal line is concentric with the outer radius.

terns. Extensions to full nonlinear analysis depend, of course, on the detailed kinetic models assumed, and can lead to persistence of an established mode beyond the size at which the linear equations predict its decay, and to bifurcation of successive modes from spatially inhomogeneous states, rather than from the homogeneous state.

Regardless of the details of the nonlinear models one might construct, the most general feature to abstract from this class of models at this stage is that, if such a dynamical system is embedded in a tissue that undergoes size and shape changes, or some other parameters such as diffusion constants change, the inevitable consequence is that a sequence of differently shaped patterns of the same morphogens arise and decay in succession at a critical sequence of bifurcation values of the size, shape, and other parameters. In particular, for an appropriate sequence of elliptical shapes, the succession of chemical patterns have nodal lines as given in Fig. 5a–f.

The set of nodal lines in Fig. 5a–f provide a model of the sequence of compartmental lines in Fig. 4b–d. The first observed line, the anterior–posterior line, arises on the blastoderm prior to the existence of the wing disc as a separate entity, and must be explained in the context of the ellipsoidal egg. Therefore, the second chemical pattern (Fig. 5b) repeats a previously formed compartmental boundary as does the fourth pattern (Fig. 5d). These correspond to the Mathieu functions Se_1 and Se_2 and follow the formation respectively of Ce_1 and Ce_2 at slightly larger sizes for "wide" ellipses.[28,29] Se_1 and Se_2 can therefore be suppressed by non-linearities in the kinetics by the prior establishment of Ce_1 (Fig. 4a) and Ce_2 (Fig. 4c). With such suppression, the total sequence and symmetry (Fig. 5g) closely fits the observed (Fig. 4c).

One means of testing the successive eigenfunction model is to assess its predictions on other imaginal discs with well-defined, but distinct shape histories. The haltere disc is similar in shape to the wing disc, but about 0.25 its size. The sequential eigenfunction model therefore predicts that only the first few patterns and compartmental lines arise. In fact, the haltere disc displays only the first three of the five lines seen on the wing disc (Fig. 4a and 4b.[27]) Figure 8a–c shows the predicted compartmental lines on the haltere, leg, and genital discs. The observed compartmental lines on the haltere, leg,[41] and genital[42] discs are fit moderately well by the predictions.

The most intriguing predictions of the model come from its application to the developing egg. The egg does not increase in size; therefore, size cannot be used as a bifurcation parameter to select different chemical patterns. However, as the nuclei migrate to the cortex of the egg, the cortical layer becomes viscous cytoplasm, rather than a relatively less

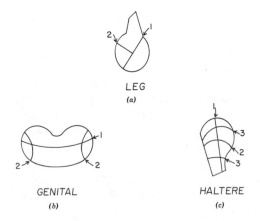

Fig. 8.   Schematic compartmental lines on the (a) leg, (b) genital, and (c) haltere.

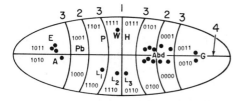

Fig. 9. Successive compartmental lines predicted on the fate map[6,7] of the blastoderm by the chemical wave model, with the binary combinatorial code assignment in each compartment generated by the successive lines. A, antenna; E, eye; Pb, proboscis; P, prothorax; W, wing and mesothorax; H, haltere; L1, L2, L3, first, second, and third legs; Abd, abdominal segments; G, genital.

viscous yolky material. It seems reasonable that diffusion constants become progressively smaller in this layer. Tuning diffusion constants smaller while holding their ratios unchanged is equivalent to tuning the amplifiable wavelengths shorter in a constant ·size physical domain. A succession of chemical instabilities will occur on the ellipsoidal egg whose patterns are the eigenfunctions of the Laplacian on that geometry.

Figure 9 shows the expectations as the amplifiable chemical wavelength gradually shortens. A first mode has a maximum at one pole and a minimum at the opposite, with a first nodal line at mid-egg dividing the egg into anterior and posterior halves. A second mode, analogous to Ce_2, with minima at the poles and a maximum in the mid-egg region, has two nodal lines, creating middle versus end egg distinctions analogous to the pair of compartmental lines that isolate the mid-disc wing region from the flanking thoracic ends (Fig. 4c). A third chemical pattern creates further longitudinal subdivisions. Finally, the amplifiable range of wavelengths shortens sufficiently to fit onto the egg from dorsal to ventral, creating a fourth compartmental line along the egg's equator.

The sequential eigenfunction model creates a sequence of different chemical patterns on a defined geometry, as parameters change. At each bifurcation value allowing a new pattern to emerge, two patterns are allowed, which are inverses, peaks replacing troughs. To select a specific pattern at each choice point requires some asymmetries in the initial conditions. Therefore, a minimum added requirement of the model is such slight chemical asymmetries in the egg and the discs.

When initially formulated, this predicted sequence of compartmental boundaries on the egg was largely hypothetical. The data then available showed that at the blastoderm stage, compartmental boundaries appeared to isolate adjacent segments, but not the dorsal from ventral disc derivatives from the three thoracic segments, since genetically marked clones at that stage can straddle from mesothorax to mesothoracic leg.[41,43,44] A short time later,[44] clonal restriction lines isolate the dorsal from ventral thoracic discs; thus, longitudinal lines of clonal restriction do arise on the egg before convincing dorsal ventral lines. Since the formulation of this model, we have tested the prediction that a first compartmental boundary line should arise at mid-egg, and confirmed the prediction.[45] The sequence in which the remaining longitudinal lines are established cannot be resolved by current techniques.

III. THE EPIGENETIC CODE

In the first sections of this article I described both the fate map of the blastoderm and the metaplasias of transdetermination and homeotic mutants. There the dominant puzzle is to understand the sense in which discs that transform into one another are program neighbors, and simultaneously to understand why that sense of "neighboring" does not map simply onto the blastoderm fate map. The transdetermination flow diagram (Fig. 3) suggests that one disc, such as genital, is simultaneously close to antenna and leg, but farther from wing. The existence of allowed and forbidden single-step transitions, sequences of transdeterminations, and the global orientation of frequencies toward mesothorax as the "sink" suggest that the determined state in each disc comprises a combination of states of independent subentities. For example, were there two genetic systems, each with two alternative dynamical steady states, 0 and 1, then genital might be encoded (00), leg (01), antenna (10), and wing (11). This encoding predicts allowed one step transitions from genital to antenna and leg, and a sequence from genital toward wing. If the 0-1 transition is easier than the reverse, a global ordering toward wing occurs.[14]

In Fig. 9, this general concept is applied to the predictions of the eigenfunction model on the egg. The first mid-egg compartmental boundary creates anterior and posterior domains, encoded with a 1 or 0 by the action of a first master gene system. The second pair of compartmental boundaries creates an ends versus middle commitment, encoded by a second master gene system in state 0 in the ends, and 1 in the middle region. The third chemical pattern creates alternating 0 and 1 assignments, subdividing existing compartments. The final dorsal–ventral equatorial line assigns a fourth master gene system a 1 state dorsally and 0 ventrally. Two dominant features of this hypothesis should be noted. First, each terminal compartment is assigned a binary combinatorial code word reflecting the sequence of alternative decisions taken during its specification. These combinations supply an epigenetic code in which each terminal compartment, corresponding to a distinctly determined progenitor zone, has a specific binary combinatorial code word. Second, because the successive chemical patterns are not monotonic on the egg, the resulting code is not monotonic on the egg; that is, in passing from anterior to posterior, the number of 0's in successive code words do not increase or decrease monotonically. This feature has the consequence that domains at opposite ends of the egg can have neighboring epigenetic code words. For example, genital (0010) and antenna (1010) differ only in the state of the first master gene system reflecting the anterior versus posterior commitment. It is this feature that shall allow this combinatorial code to account for the intriguing patterns of metaplasia seen in transdetermination and homeotic mutants.

Predictions with respect to transdetermination are made on the following basis: the conversion of haltere to wing (Fig. 9) requires conversion of the first "switch" from 0 to 1. Conversion of haltere to antenna requires the same conversion of this first "switch," but also conversions of the second and fourth master gene systems from 1 to 0; hence, transdetermination of haltere to wing should occur more frequently than haltere to antenna. Figure 3 confirms this prediction. Table I lists 37 independent predictions of this coding model. Almost all are true, two or perhaps three are false. The *a priori* probabilities of such a success rate are very small, and suggest that this combinatorial scheme accounts well for the sense in which developmental programs in different discs are neighbors with respect to transdetermination.

The same combinatorial code accounts well for a broad spectrum of homeotic conversions, including those that cause conversion of tissues

TABLE I. Predicted Relative Transdetermination Frequencies Derived from the Chemical Wave Model Applied to the Blastoderm[a]

Prediction	Status	Preduction	Status	Prediction	Status
$H \to W > H \to A$	T	$A \to W > A \to H$	T	$L \to W > L \to E$	T
$H \to W > H \to L_{1,2}$	T	$A \to L > A \to W$	F	$L_{1,2} \to W > L_{1,2} \to H$	T
$H \to W > H \to E$	T	$A \to Pb > G \to Pb$	?	$L \to A > L \to E$	T
$H \to W > H \to Pb$	T	$A \to E > A \to W$	F	$L_{1,2} \to A > L_{1,2} \to G$	T
$W \to A > H \to A$	T	$A \to G > L_{1,2} \to G$	T	$L_2 \to G > L_3 \to A$	?
$W \to E > H \to E$	T	$A \to E > E \to A$	T	$L_1 \to Pb > L_1 \to G$	?
$W \to L_{1,2} > H \to L_{1,2}$	T	$A \to L_2 > L_2 \to A$	?T	$G \to A > G \to Pb$	T
$W \to L > W \to A$	T	$E \to W > E \to H$	T	$G \to A > G \to W$	T
$W \to L > W \to G$	T	$E \to A > E \to G$	T	$G \to L_{2,3} > G \to W$	?T
$W \to A > W \to G$	T	$E \to A > E \to L$	T	$G \to A > G \to L_{1,2}$	?T
$W \to E > W \to Pb$	T	$E \to W > E \to L$	T	$G \to A > A \to G$	T
$W \to E > W \to G$	T			$G \to L > L \to G$	T
$W \to E > W \to A$	?			$G \to H > G \to W$	?F

[a] $L_{1,2} \to A > L_{1,2} \to G$ means the model predicts transdetermination from the first or second leg to antenna is greater than to genital. Abbreviations are explained in the legend of Fig. 9.

from one end of the fate map to the other, such as *tumorous head* which converts head to genital and abdominal tissue. If a specific homeotic mutant alters the state of a single master gene then the following three general properties should occur: (1) Some homeotic transformations should occur between nonneighboring domains of the fate map. (2) If the same mutant alters the same master gene in two different tissues, the two should be transformed to two distinct new tissues, that is, coordinated *parallel* homeotic transformations should occur. This is exemplified by *Nasobemia* which converts antenna to mesothoracic leg, and eye to wing.[16] (3) If most homeotic mutants act on single master genes, but transdetermination allows "wobble" of any decision, then the set of tissues to which any disc can transdetermine should be broader than, but inclusive of, those to which any specific homeotic mutant transforms it. All these general ordering relations obtain (see review by Ouweneel[15]). If the specific combinatorial code in Fig. 9 is a good model, it should account for the actual neighboring relations between discs seen in homeotic mutants. That is, most such transformations should be accounted for by alterations in a single binary digit in the code. Table II shows that it is fair to say that

TABLE II

Mutant	Symbol	Transformation		Coordination	Code Change	Switches Required
Antennapedia[1]	Antp	antenna	→ leg 2	-	1010→1110	1
Pointed wing	Pw	antenna	→ wing	-	1010→1111	2
Nasobemia	Ns	antenna	→ leg 2	parallel	1010→1110	1
		eye	→ wing		1011→1111	1
dachsous	ds	tarsus	→ arista	-	1110→1010	1
Opthalmoptera[2]	OptG	eye	→ wing	-	1011→1111	1
Hexaptera	Hx	prothorax	→ mesothorax	-	1101→1111	1
podoptera	pod	wing	→ leg	-	1111→1110	1
tetraltera[3]	tet	wing	→ haltere	-	1111→0111	1
Contrabithorax	Cbx	wing	→ haltere	parallel	1111→0111	1
		leg 2	→ leg 3		1110→0110	1
Ultrabithorax	Ubx	haltere	→ wing	parallel	0111→1111	1
		leg 3	→ leg 2		0110→1110	1
tumorous head	tuh1,3	eye	→ genital	parallel	1011→0011	1
		antenna	→ genital	divergent	1010→0010	1
		antenna	→ leg		1010→1110	1*
lethal(3)III-10	1(3)III-10	haltere	→ wing	parallel	0111→1111	1
lethal(3)XVI-18	1(3)XVI-18	genital	→ antenna	divergent	0010→1010	1
		genital	→ leg		0010→0110	1*
lethal(3)703	1(3)703	antenna	→ leg	parallel	1010→1110	1
lethal(3)1803R	1(3)1803R	genital	→ leg	divergent	0010→0110	1
		genital	→ antenna	parallel	0010→1010	1*
		haltere	→ wing		0111→1111	1
proboscipedia	pb	proboscis	→ antenna	divergent	1000→1010	1
		proboscis	→ leg		1000→1100	1*
extrasexcombs[4]	ecs	leg 2	→ leg 1	convergent	1110→1100	1
		leg 3	→ leg 1		0110→1100	2
Polycomb	Pc	antenna	→ leg 2		1010→1110	1
lethal(4)29	1(4)29	leg 2	→ leg 1	convergent	1110→1100	1*
		leg 3	→ leg 1		0110→1100	2*

this combinatorial scheme accounts for the neighboring relations of developmental programs intimated by both homeosis and transdetermination, and relates it with moderate success to the geometry of the blastoderm fate map.

It would be astonishing were this particular scheme to be correct in detail. The existence of convergent and divergent homeotic mutants shows that it probably is not. If it is assumed that a single homeotic mutant can affect at most a single master gene system, then a binary code model precludes either convergence or divergence. It is in this sense that the existence of these classes of homeotic mutants suggests that different pictures of the underlying genetic logic must be assumed to provide minimal models of homeotic transformation. For example, a variety of mutants transforms second and third legs into first legs. Both the second and third leg cannot differ from one another and from the first leg by the binary state of a single gene, or genetic system. At a minimum, a genetic system with three alternative states must be assumed, coupled with the assumption that the mutant converts two of these states to the third. Because convergent and divergent homeotic mutants are minority classes among homeotics, and require more complex logic, it may prove useful to assume that those exhibiting either single transformation or parallel transformations are the general class, and the convergent and divergent classes special aberrations from the norm. Then the general class can in principle be accounted for by a binary combinatorial code. Three state genetic systems can be obtained in several ways; one simple way is an appropriate coupling of two genetic circuits with two states each, to yield a three-state device.

Since its formulation, a number of interesting mutants bearing on both the eigenfunction model and combinatorial model have been discovered. The successive eigenfunction model proposes that ever-shorter wavelength patterns successively divide up the egg into smaller subdomains. Mutants of such a system might be expected to produce embryos that failed to undergo the complete sequence. Recently, a larval lethal mutant has been discovered which yields an embryo having six wide longitudinal segments rather than 12 narrow segments.[46] It appears that final partition of these six, each into two segments, has not occurred. Mirror symmetrical aberrations have now been discovered in a remarkable spectrum of spatial lengths on the *Drosophila* embryo. Well known in both *Drosophila* and *Smittia* are mutants and experimental procedures that produce double abdomen embryos with mirror symmetry about an anterior abdominal or metathoracic segment.[12,47] In these cases, the mirror symmetry extends over half the embryo. It is attractive to account for such transformations by assuming that the first master gene in Fig. 9, which records the anterior–posterior decision, is left uniformly in the posterior 0 state, yielding a mirror symmetric double abdomen (see also Kalthoff[12]). Were the second master gene system, distinguishing ends from middle, to mutate to a constitutive 0 state, the embryo would consist of mirror symmetric posterior abdominal segments in the abdominal half of the embryo and mirror symmetric head segments in the anterior half of the embryo. A recent early lethal mutant, still only partially characterized,[48] has mirror symmetric posterior abdominal segments in the posterior half of the embryo, is missing the thoracic segments, and has grossly abnormal head structures whose details are difficult to elucidate because head segments are inverted inside the hatching larva. In the posterior half embryo, where mirror symmetry is clear, it ranges over one-fourth the length of the body plan. Another lethal mutant yields an embryo with four large segments, each bearing mirror symmetrical arrangements of ventral denticles.[48] Here, mirror symmetry ranges locally over $\frac{1}{8}$ the embryo length. Failure of the third digit in Fig. 9 to function normally, such that it remained constitutively in the 0 state would yield four wide domains. Mirror symmetry within some or all of these domains would be obtained if an additional longitudinal chemical pattern normally subdivided the eight segments of Fig. 9 into the full 12 segments of the embryo, analogous to the mirror symmetries obtained if the second digit is constitutively 0 in Fig. 9. Finally a mutant has been uncovered which is internally mirror symmetric in the cuticular patterns of all larval thoracic and abdominal segments.[48]

This broad spectrum of embryonic lengths over which mirror symmetry occurs is clearly suggestive of a variety of chemical patterns of successively shorter wavelengths, with successively more peaks and troughs, fitting onto the egg. Although provocative at this stage, it is obvious that the class of models I have discussed remains a macroscopic analogy. It should be stressed that some of the mutants are still incompletely characterized, and that a variety of other mechanisms might account for the patterns observed.

IV.   DISCUSSION

Reaction-diffusion systems provide a means to subdivide successively a domain at a sequence of critical parameter values due to size, shape, diffusion constants, or other parameters. The chemical patterns that arise are the eigenfunctions of the Laplacian operator on that geometry. The succession of eigenfunctions on geometries close to the wing, leg, haltere, and genital discs yield sequential nodal lines reasonably similar to the observed sequence and symmetries and geometries of the observed compartmental lines. The identity is not perfect. The hypothesis that the egg is subdivided by such sequential eigenfunctions into subdomains yields a combinatorial epigenetic code, which interprets a wide variety of homeotic mutants and transdetermination relations as transitions between neighboring developmental code words. The accord is good, but not perfect. Recent experiments and mutants on the egg confirm that an initial anterior–posterior compartmental boundary exists, that longitudinal compartments arise successively, and prior to a major dorsal-ventral line in the thoracic regions. Mutants reveal the existence of embryos with half the number of normal segments, and mirror symmetrical segmental structures whose range varies from $\frac{1}{2}$ egg, $\frac{1}{4}$ egg, $\frac{1}{8}$ egg, to mirror symmetry within individual longitudinal segments. These patterns at least suggest chemical gradients with varying numbers of longitudinal peaks and troughs.

Acknowledgments

This research has been supported in part by grants GM 22341-04 from the National Institutes of Health, PCM 78-15337 from the National Science Foundation, and CD-30 from the American Cancer Society.

APPENDIX

This section deals with the analysis of reaction-diffusion systems.*

In a chemical system, spatial patterns can spontaneously arise from an initial spatially homogeneous concentration profile by the selection from noise and amplification of perturbations with wavelengths in the neighborhood of some preferred wavelength. To determine the conditions under which this occurs, the behavior of the system in the vicinity of the spatially homogeneous steady state $X = X_s$, $Y = Y_s$, where $F(X_s, Y_s) = G(X_s, Y_s) = 0$, is analyzed using a standard linearization procedure.

The system is linearized about the spatially homogeneous steady state by substituting $X(r, t) = X_s + x(r, t)$, $Y(r, t) = Y_s + y(r, t)$. and retaining only terms up to first order in x and y in a Taylor expansion of $F(X, Y)$ and $G(X, Y)$. The resulting linear equations in x and y are

$$\frac{\partial x}{\partial t} = K_{11}x + K_{12}y + D_1\nabla^2 x$$

$$\frac{\partial y}{\partial t} = K_{21}x + K_{22}y + D_2\nabla^2 y$$

(A1)

* From S. Kauffman, R. Shymko, and K. Trabert, "Control of Sequential Compartment Formation in *Drosophila*," *Lectures on Mathematics in the Life Sciences* **10**, 207–243 (1978).

These equations are solved by separating out the time dependence through the substitutions $x(r, t) = x'(r)e^{\lambda t}$, $y(r, t) = y'(r)e^{\lambda t}$, and diagonalizing the resulting pair of spatially dependent coupled equations. These two separated equations are Helmholtz-type equations whose solutions can be straightforwardly obtained in different coordinate systems.[28,49] The complete space-time-dependent solutions are sums of spatial modes or patterns, each with a characteristic temporal behavior. For example, the complete solution on a circle can be written

$$\begin{bmatrix} x(r, \phi, t) \\ y(r, \phi, t) \end{bmatrix} = \sum_{i=+,-} \sum_{j=1}^{\infty} \sum_{n=0}^{\infty} \begin{bmatrix} a_{nji} \\ b_{nji} \end{bmatrix} \times \exp \lambda_{nji} t J_n(k_{nj}r) \cos n\phi \quad \text{(A2)}$$

where J_n is the nth Bessel function and r and ϕ are the radial and angular coordinates, respectively, a_{nji} and b_{nji} are arbitrary constants to be determined by initial conditions. Each mode has two temporal eigenvalues, indicated by the sum $i = +, -$. The sum over j depends on the boundary conditions chosen.

For no-flux boundary conditions, the spatial gradient at the boundary must have zero component normal to the boundary.[49] In a circle of radius r_0, this means that $\partial x(r, o, t)/\partial r = \partial y(r, o, t)/\partial r = 0$ at $r = r_0$. The zeros in the derivatives of $J_n(z)$ occur at particular values of the argument $z = z_{nj}$.[50] Therefore, the spatial mode $J_n(k_{nj}r) \cos n\phi$, which we abbreviate by J_{nj}, is obtained when the jth zero in the derivative of J_n occurs at the boundary; that is, when $k_{nj}r_0 = z_{nj}$. This fixes the value of k_{nj} associated with the mode J_{nj} for any given radius r_0. As the radius changes, the value of k_{nj} changes in inverse proportion.

The temporal behavior of the mode J_{nj} is determined by the dynamics through the dispersion relation between the temporal eigenvalues λ_{nji} and the spatial eigenvalue k_{nj}. For equations (A1) this relation has the form

$$\lambda_{nj\pm} = \tfrac{1}{2}[K_{11} + K_{22} - k_{nj}^2(D_1 + D_2) \pm \{[K_{11}$$
$$- K_{22} - k_{nj}^2(D_1 - D_2)]^2 + 4K_{12}K_{21}\}^{1/2}] \quad \text{(A3)}$$

As seen in equation (A2), each mode J_{nj} behaves in time according to a sum of terms of the form $A \exp(\lambda_{nj+}t) + B \exp(\lambda_{nj-}t)$, where A and B are specified by initial conditions. Therefore, any mode with the real parts of λ_{nj+} and λ_{nj-} both negative will decay and disappear; a mode with either or both real parts positive will grow and create a spatial pattern.

On an ellipse, the solutions of equation (A1) are

$$\begin{bmatrix} x(\xi, \eta, t) \\ y(\xi, \eta, t) \end{bmatrix} = \sum_{i=+,-} \sum_{j=1}^{\infty} \sum_{n=0}^{\infty} \begin{bmatrix} a_{nji} \\ b_{nji} \end{bmatrix}$$

$$\times \exp \lambda_{nji} t [Ce_n(\xi, s_{nj})ce_n(\eta, s_{nj}) + S_{nji}Se_n(\xi, s_{nj})se_n(\eta, s_{nj})] \quad \text{(A4)}$$

Here ce_n and se_n are periodic cosine- and sine-elliptic Mathieu functions, respectively, of integral order and Ce_n and Se_n are the corresponding nonperiodic (or modified) Mathieu functions.[28] ξ and η are the elliptical coordinates tracing out confocal ellipses and hyperbolae, respectively (Fig. 4), and $s_{nj} = h^2 k_{nj}^2$, where h is one-half the interfocal distance of the ellipse. The constants a_{nji}, b_{nji}, and S_{nji} are determined by initial conditions. We use the abbreviations Ce_{nj} or Se_{nj} for the patterns $Ce_n(\xi, s_{nj})ce_n(\eta, s_{nj})$ or $Se_n(\xi, s_{nj})se_n(\eta, s_{nj})$, respectively.

At the boundary of the ellipse, $\xi = \xi_0$, the no-flux condition becomes $\partial Ce_n(\xi, s_{nj})/\partial\xi = 0$ or $\partial Se_n(\xi, s_{nj})/\partial\xi = 0$, analogous to the circular case. This condition fixes the value of the scaling factor k_{nj} for the pattern Ce_{nj} or Se_{nj}, and the temporal behavior of this pattern is determined by exactly the same dispersion relation [equation (A3)] as for the circular case. Those modes which grow in time will form spatial patterns on the ellipse; those which decay in time will not be seen.

If the following five conditions hold

$$\text{(i)} \quad K_{11} + K_{22} < 0$$

$$\text{(ii)} \quad K_{11}K_{22} - K_{12}K_{21} > 0$$

$$\text{(iii)} \quad (K_{11} - K_{22})^2 > -4K_{12}K_{21} \quad \text{(A5)}$$

$$\text{(iv)} \quad D_1 K_{22} + D_2 K_{11} > 0$$

$$\text{(v)} \quad \left(\sqrt{\frac{D_1}{D_2}} K_{22} - \sqrt{\frac{D_2}{D_1}} K_{11} \right)^2 > -4K_{12}K_{21}$$

spatial patterns will arise spontaneously. In such a system, one reactant diffuses more readily than the other, one catalyzes its own production while the other inhibits its own production, and one catalyzes the production of the second while the second inhibits the production of the first. Under these conditions, the temporal eigenvalues λ_{nj+} and λ_{nj-} will both be real and the larger of them, λ_{nj+}, will be positive only in the neighborhood of a particular value of k_{nj}, equal to $2\pi/l_0$ where l_0 is the natural chemical wavelength of the system. Only those modes with positive λ_{nj+} will grow which satisfy the no-flux boundary condition with $k_{nj} \approx 2\pi/l_0$. Furthermore, since λ_{nj+} is real, this pattern will grow without oscillation.

For an ellipse with a given eccentricity $\epsilon = 1/\cosh \xi_0$, the wave number k_{nj} for the pattern Ce_{nj} or Se_{nj} is inversely proportional to the interfocal distance, $2h$, and therefore to the size of the ellipse,[28] in direct analogy with the circular case. Therefore as the size increases, each k_{nj} will be scaled downward along the abscissa, and different modes will appear in sequence as their respective k_{nj}'s enter the region of positive λ_{nj+}. However, since more and more modes will be compressed into the region of small k_{nj}, eventually more than one mode will fall in the positive λ_{nj+} region. In this case, a superposition of modes might appear or, in a fully nonlinear system, a previously established mode might suppress a later mode even though both are allowed in the linear theory.

In the linear approximation given by equations (A1), modes selected for amplification grow without bound. The nonlinear reaction-diffusion system

$$\frac{\partial X}{\partial t} = -AX + \frac{BY^n}{1 + Y^n} + D_1 \nabla^2 X$$

$$\frac{\partial Y}{\partial t} = -CX + \frac{D(Y^n + b)}{1 + Y^n} + D_2 \nabla^2 Y \tag{A6}$$

is an example of a system which can create spatial patterns as described above, but in which these patterns grow only to a finite size. In this system, X inhibits both its own and Y's production, and Y catalyzes the production of both. Also, X diffuses more readily than Y.

In equations (A6), with $A = 7.8$, $B = 15.6$, $C = 1$, $D = 1.7$, $b = 0.2$, $n = 6$, $D_1 = 16$, $D_2 = 1$, this system has one spatially homogeneous steady state at $X_s = Y_s = 1$. Linearizing about this steady state and substituting the resulting linearization constants K_{ij} into the dispersion relation [equation (A3)] we find that λ_{nj+} is positive for k_{nj} between 0.7 and 1.0, corresponding to wavelengths between $l_{\min} = 6.1$ and $l_{\max} = 9.1$. In a one-dimensional domain $0 < r < L$, the linearization of equation (A6) would therefore amplify the pattern $\cos n\pi r/L$ henceforth called the "n-mode," in the range of lengths $\frac{1}{2}nl_{\min} < L < \frac{1}{2}nl_{\max}$. Computer simulations showed that the analogous nonlinear patterns appeared in the identical length ranges, although their shapes were slight distortions of pure cosines.

References

1. W. Gehring and R. Nöthiger, The imaginal discs of *Drosophila*. In *Developmental Systems: Insects*, Vol. 1, S. J. Counce and C. H. Waddington, eds., Academic Press, New York, 1973, pp. 212–290.

2. F. R. Turner and A. P. Mahowald, Scanning electron microscopy of *Drosophila* embryogenesis. I. The structure of the egg envelope and the formation of the cellular blastoderm. *Dev. Biol., 50*, 95–103 (1976).

3. F. R. Turner and A. P. Mahowald, Scanning electron microscopy of *Drosophila melanogaster* embryogenesis. II. Gastrulation and segmentation. *Dev. Biol., 57*, 403–416 (1977).

4. R. Nöthiger, The larval development of imaginal discs. In *Results and Problems in Cell Differentiation*, V. C. H. Ursprung and R. Nöthiger, eds., Springer-Verlag, Berlin, 1972, pp. 1–34.

5. E. Hadorn, Dynamics of determination. *Symp. Soc. Dev. Biol., 25*, 85–104 (1966).

6. Y. Hotta and S. Benzer, Mapping behavior in *Drosophila* mosaics. *Nature (London), 240*, 527–535 (1972).

7. A. Garcia-Bellido and J. R. Merriam, Cell lineage of the imaginal discs in *Drosophila* gynandromorphs. *J. Exp. Zool., 170*, 61–76 (1969).

8. W. Janning, Gynandromorph fate maps in *Drosophila*. In *Genetic Mosaics and Cell Differentiation*, W. J. Gehring, ed., Springer-Verlag, Berlin, 1978, pp. 1–28.

9. L. Wolpert, Positional information and pattern formation. In *Current Topics in Developmental Biology*, A. A. Moscana and A. Monroy, eds., 1971, Vol. 6, pp. 183–224.

10. K. Sander, Pattern specification in the insect embryo. In *Cell Patterning*. Ciba Foundation Symposium 29, Elsevier, Amsterdam, 1975, pp. 241–263.

11. K. Sander, Current understanding of cytoplasmic control centers. In *Insect Embryology*, S. W. Visscher, ed., Montana State University Press, Bozeman, 1977, pp. 31–61.

12. K. Kalthoff, Analysis of a morphogenetic determinant in an insect embryo (*Smittia* sp., Chironomidae, Diptera). In *Determinants of Spatial Organization*, S. Subtelny and I. Konigsberg, eds., Academic Press, New York, 1979, pp. 97–126.

13. S. A. Kauffman, The compartmental and combinatorial code hypotheses in *Drosophila* development. *Bioscience, 29*, 581–588 (1979).

14. S. A. Kauffman, Control circuits for determination and transdetermination. *Science, 181*, 310–318 (1973).

15. W. I. Ouweneel, Developmental genetics and homeosis. *Adv. Genet., 18*, 179–248 (1976).

16. V. P. Stepshin and E. K. Ginter, A study of the homeotic genes *Antennopedia* and *Nasobemia* of *Drosophila melanogaster*. *Genetika, 8*, 93 (1972).

17. J. Haynie, 21st Annual *Drosophila* Conference, 1980.

18. J. H. Postlethwait, P. Bryant, and G. Schubiger, The homeotic effect of "tumorous head" in *Drosophila melanogaster Dev. Biol., 29*, 337–342 (1972).

19. E. B. Lewis, Genetic control and regulation of developmental pathways. In *The Role of Chromosomes in Development*, M. Locke, ed., 23rd Symposium of the Society for the study of Development and Growth, Academic Press, New York, 1964, pp. 231–251.

20. E. B. Lewis, A gene complex controlling segmentation in *Drosophila*. *Nature (London)*, **276**, 565–570 (1978).

21. S. A. Kauffman, R. M. Shymko, and K. Trabert, Control of sequential compartment formation in *Drosophila*. *Science*, **199**, 259–270 (1978).

22. A. Hannah-Alava, Developmental genetics of the posterior legs in *Drosophila melanogaster*. *Genetics*, **43**, 878–905 (1958).

23. A. Garcia-Bellido, P. Ripoll, and G. Morata, Developmental compartmentalization of the wing disk of *Drosophila*. *Nature New Biol.*, **245**, 251–253 (1973).

24. H. J. Becker, Mitotic recombination and position effect variegation. In *Genetic Mosaics and Cell Differentiation*, W. Gehring, ed., Springer-Verlag, New York, 1978, pp. 29–49.

25. J. R. Merriam, Estimating primordial cell numbers in *Drosophila* imaginal discs and histoblasts. In *Genetic Mosaics and Cell Differentiation*, W. Gehring, ed., Springer-Verlag, New York, 1978, pp. 71–96.

26. P. Bryant, Pattern formation in the imaginal wing disc of *Drosophila melanogaster*: fate map, regeneration and duplication. *J. Exp. Zool.*, **193**, 49–77 (1975).

27. A. Garcia-Bellido, Genetic control of wing disc development in *Drosophila*. In *Cell Patterning*, Ciba Foundation Symposium 29, Elsevier, Amsterdam, 1975, pp. 161–182.

28. N. W. McLachlan, *Theory and Applications of Mathieu Functions*, Clarendon, Oxford, 1947.

29. M. J. King and J. C. Wiltse, Derivative zeros and other data pertaining to Mathieu functions. Johns Hopkins Radiation Laboratory Technical Report No. AF57, Baltimore, Maryland, 1958.

30. R. G. Harrison, On the relations of symmetry in transplanted limbs. *J. Exp. Zool.*, **32**, 1–136 (1921).

31. P. A. Lawrence, The development of spatial patterns in the integument of insects. Pages 157–209 In *Developmental Systems: Insects, Vol. 2*, S. J. Counce and C. H. Waddington, eds., Academic Press, New York, 1973, pp. 157–209.

32. V. French, P. J. Bryant, and S. V. Bryant, Pattern regulation in epimorphic fields. *Science*, **193**, 969–981 (1976).

33. B. C. Goodwin, and L. E. H. Trainor, A field description of the cleavage process in embryogeneis. *J. Theor. Biol.*, **85**, 757–770 (1980).

34. H. Meinhardt, A model of pattern formation in insect embryogenesis. *J. Cell. Sci.*, **23**, 117–139 (1977).

35. S. A. Kauffman, Pattern generation and regeneration. In *Pattern Formation*, G. M. Malacinski, ed., Macmillan, New York, 1983.

36. M. Hershkowitz-Kaufman, Bifurcation analysis of nonlinear reaction-diffusion equations. II. Steady state solutions and comparison with numerical simulations. *Bull. Math. Biol.*, **37**, 589–636 (1975).

37. A. M. Turing, *Philos. Trans. R. Soc. London Ser. B*, **237**, 37 (1952).

38. J. I. Gmitro and L. E. Scivin, In *Intracellular Transport*, K. B. Warren, ed., Academic Press, New York, 1966.

39. G. Nicolis and I. Prigogine, *Self-Organization in Nonequilibrium Systems*, Wiley-Interscience, New York, 1977.

40. A. Babloyantz and J. Hiernaux, Models for cell differentiation and generation of polarity in diffusion-governed morphogenetic fields. *Bull. Math. Biol.*, **37**, 637–657 (1975).

41. E. Steiner, Establishment of compartments in the developing leg imaginal discs of *Drosophila melanogaster. Wilhelm Roux's Arch. Dev. Biol.,* **180,** 9–30 (1976).

42. K. Dubendorfer, Unpublished Ph.D. thesis, University of Zurich, 1977.

43. E. Wieschaus and W. Gehring, Clonal analysis of primordial disc cells in the early embryo of *Drosophila melanogaster. Dev. Biol.,* **50,** 249–265 (1975).

44. P. Lawrence and G. Morata, The early development of mesothoracic compartments in *Drosophila. Dev. Biol.,* **56,** 40–51 (1977).

45. S. A. Kauffman, Pattern formation in the *Drosophila* embryo, *Phil. Trans. Roy. Soc. London B,***295,** 567–594 (1981).

46. T. Kaufman, 21st Annual *Drosophila* Conference 1980.

47. C. Nüsslein-Volhard, Maternal effect mutations that alter the spatial coordinates for the embryo of *Drosophila melanogaster.* In *Determinants of Spatial Organization,* S. Subtelny and I. Konigsberg, eds., Academic Press, New York, 1979, pp. 185–211.

48. E. Wieschaus, personal communication, 1980.

49. P. M. Morse, and H. Feshbach, *Methods in Theoretical Physics,* McGraw-Hill, New York, 1953.

50. M. Abramowitz, and I. A. Stegun, *Handbook of Mathematical Functions,* Government Printing Office, Washington, D.C., 1972.

LOGICAL DESCRIPTION, ANALYSIS, AND SYNTHESIS OF BIOLOGICAL AND OTHER NETWORKS COMPRISING FEEDBACK LOOPS

R. THOMAS

Department of Molecular Biology
University of Brussels, Belgium

The types of regulation observed in biological systems and their relation with systems in other fields are discussed in Section I. *Homeostatic* regulations tend to keep a rate of operation near a supposedly optimal level, usually well below the level that would be observed in the absence of regulation. *Epigenetic* regulations endow systems with the possibility of "choosing" between two or more stable states of regime. Clearly, homeostatic regulations are related to the mechanisms, studied by theoretical physicists, that can generate sustained oscillations, and epigenetic regulations are related to the mechanisms that can generate multiple steady states. All these behaviors imply *feedback loops* in the logical structure of the systems. We recall the distinction between *negative* and *positive* feedback loops, which are characterized, respectively, by an odd and an even number of negative (inhibitory) interactions. Negative and positive feedback loops account, respectively, for homeostatic and epigenetic types of regulation.

Most systems involve several interconnected feedback loops. Such systems cannot be analyzed seriously without a proper formalism, but their detailed description using differential equations is often too heavy. For these reasons we (as many others before) turned to a *logical* (or Boolean) description, that is, a description in which variables and functions can take only a limited number of values, typically two (1 and 0). Section II is an updated description of a logical method ("kinetic logic") whose essential aspects were first presented by Thomas[1] and Thomas and Van Ham.[2] A less detailed version of this part can be found in Thomas.[3] The present paper puts special emphasis on the fact that for each system the Boolean trajectories and final states can be obtained *analytically* (i.e.,

247

without ascribing a numerical value to the time delays), and in particular on *logical stability analysis*.

Section III discusses briefly (1) the relation between our logical equations and the differential equations as used in chemical kinetics; (2) some aspects of logical versus numerical iteration methods; (3) the possible application of our method (initially developed for genetic purposes) to other fields, and more particularly chemical kinetics; and (4) the possibility of using this method *al rovescio*, that is, in a *synthetic* (*inductive*) way. In this perspective we assume that the essential elements of a system have been correctly identified and we ask to what extent one can proceed rationally from the observed behavior toward sets of interactions which account for this behavior.

I. REGULATION IN BIOLOGICAL SYSTEMS

A. Biological Regulations: Homeostatic or Epigenetic

When one refers to biological regulations, one usually thinks of homeostatic regulations, that is, regulations that tend to maintain the level of variables in the vicinity of a supposedly optimal value, usually well below the level that would be reached if the system operated at its maximal rate. I shall discuss these mechanisms in Section I.D. Note that this homeostatic type of regulation operates essentially like a thermostat; in practice, even where the objective is constancy, the system usually oscillates around the chosen value because this regulation proceeds by alternate upward and downward corrections.

But there is another, no less essential, type of regulation, which endows biological systems with the possibility to diverge from a common initial situation and persist indefinitely in two or more distinct states of regime. One speaks of *epigenetic* differences when two cell lines are, as far as one can tell, genetically identical and yet display permanent, hereditary differences in the absence of actual environmental differences. The best-documented cases of epigenetic differences, in which the mechanisms have been really cleared up, are found among microorganisms.[4-8] But many biologists, including myself, are convinced that a large sector of *differentiation* in higher organisms has an epigenetic nature. If this view is correct, the understanding of epigenetic differences will be of utmost importance. Note that "epigenetic" should not really be *opposed* to "genetic": although an epigenetic difference between two cell lines is a hereditary difference not based on a difference of the structure of the genetic material, the very possibility of displaying different epigenetic states depends on the structure of the genetic network (and, more specifically, on the presence of positive loops in this network) (see Section I.D.).

B. Continuous and Discontinuous Descriptions

As shown notably by the thermodynamic school of Brussels,[9,10] systems maintained far from equilibrium and endowed with appropriate nonlinearities and feedback interactions may display such nontrivial behaviors as sustained oscillations and multiple steady states (see papers by I. Prigogine and G. Nicolis in this volume). Even though the structures studied are much simpler than biological systems, this type of work provides a firm fundamental basis, not sufficient but absolutely necessary, for the future understanding of such processes as cell differentiation. Clearly, the sustained oscillations and multiple steady states displayed by simpler systems are related, respectively, with the two types of biological regulation described above.

I would like to emphasize that the needs for appropriate nonlinearities and for appropriate feedback interactions are distinct conceptually. One can write a system of linear differential equations which is looped on itself, or a system comprising one or more nonlinearities which is not looped on itself. In neither case, I guess, will one obtain multiple steady states or sustained oscillations. Appropriate nonlinearities and appropriate logical loops are apparently both necessary for the nontrivial behavior. Let us consider a system composed of feedback loops and first try to describe it by a set of *linear* differential equations; no complex behavior is expected. We now introduce nonlinearities in the form of sigmoidal interactions, whose degree of nonlinearity can be characterized by the Hill coefficient n. As n increases,* the system may acquire a more complex behavior (sustained oscillations and/or multiple steady states, etc.). However, one rapidly reaches a plateau beyond which the complexity of the behavior no longer increases; from this point on, the qualitative behavior will not change, even if n tends to infinity and the sigmoid interaction tends to a step function (see Glass and Kaufmann[12] for specific examples). This is an expression of the fact that complexity is limited not only by the degree of nonlinearity but also by the logical structure of the system.

A possible strategy is thus to describe the systems considered in logical (i.e., all-or-none) terms, as if each interaction was infinitely nonlinear; on the one hand, the treatment is easy, and on the other hand, one may expect that in spite of its apparent caricatural character this description will usually be qualitatively the same as the continuous one. This attitude

---

* As shown by Richelle[11] the important point is the *overall* degree of nonlinearity of the loops. For instance, a loop composed of three sigmoid inhibitory interactions can generate sustained oscillations if one of the interactions has a Hill number ≥ 8, but also if all three interactions have a Hill number ≥ 2. In this case, what is important is the product of the Hill numbers.

amounts to asking: Granted the other requirements for "nontrivial" behaviors (removal from equilibrium, sufficient nonlinearities), how exactly does the range of behaviors of a system depend on its logical structure?

C. The Type of Systems Treated

The systems we have been treating are essentially networks whose elements influence the rate of operation of each other. For instance, in the field of genetics, each gene operates by directing the synthesis of a characteristic product, usually a protein; we are especially interested in sets of interacting genes in which the rate of operation of each gene is influenced by the concentration of one or more gene product(s). In this type of system, the elements considered are mostly products, and we consider the influence exerted by the *concentration* of the products on their *rates of synthesis*.

We felt from the beginning, and it appears to be essentially correct, that situations in many fields other than genetics, not only in biology but in other disciplines as well, are sufficiently similar from the logical viewpoint to be formalized in the same way. Of course, the very nature of the elements and of the interactions is completely different; for instance, when we describe a thermostat we consider temperatures rather than concentrations and rates of heat production rather than rates of product synthesis. More generally, we try to relate *fluxes* to *levels*.

Note that in the systems we are treating in genetics, the interactions between the elements usually act in a catalytic way: the elements influence the rate of synthesis (or the activity) of each other without being consumed as a result of this action. However, in chemistry, a "positive" influence of a substance α on the rate of synthesis of another substance β more often consists in the transformation of α into β rather than in a catalytic effect of α on the rate of β synthesis! As we will see (Section III.C), these situations require some caution in the logical treatment.

D. Feedback Loops

Systems that display such complexities as sustained oscillations or multiple steady states always comprise feedback loop(s) in their logical structure. In our context this may mean, for instance, that substance α exerts a control on the rate of synthesis of β, which exerts a control on the rate of synthesis of γ, which in turn exerts a control on the rate of α synthesis. More generally, there are elements whose rate of production is directly or indirectly influenced by their own concentration.

In a *simple* (unbranched) feedback loop, each element is under direct control *only* of the element immediately upstream in the loop and it exerts a direct control *only* on the element immediately downstream in the loop.

But it is also true that each element exerts an indirect effect on the production of all the elements of the loop, including itself.

What is perhaps not immediately obvious is that in any simple loop, whatever its number of elements and whatever the number and order of the positive and negative interactions, either *all* the elements of the loop exert a positive control on their own production, or *all* the elements of the loop exert a negative control on their own production. There are thus two classes of simple feedback loops. In the loops of one class, each element exerts indirectly, via all the other elements, a negative control on its own further production (or activity); let us call these loops *negative loops* as already suggested in Thomas et al.[13,14] In the other class, each element exerts indirectly, via all the other elements, a positive control on its own further production (or activity); let us call these loops *positive loops*.[13,14]

Whether a loop is positive or negative depends only on the parity of its number of negative interactions: a loop with an *odd* number of negative interactions is a *negative* loop, a loop with an *even* number of negative interactions is a *positive* loop.[13,14]

One may now ask: What are the respective behaviors of negative and positive loops? If α directly or indirectly inhibits its own production (negative loops), its level will tend to oscillate around an average value (or stabilize at this value) because when the level is too high the production will be blocked; when, as a delayed result of this block (and of the spontaneous decay of α) the level has become low enough, the production resumes, etc. The logical treatment (specific examples: Refs. 1 and 15; generalization: Ref. 14; formal demonstration: Ref. 16) ascribes a periodic behavior to all simple negative loops. In the continuous description, these systems may display a periodic behavior (limit cycle) only when the number of elements is ≥ 3 and the nonlinearity is sufficient. But, if one takes into account the occurrence of incompressible delays (as, e.g., the time required to synthesize a protein from a gene) a negative loop may generate sustained oscillations even if it has only one element.[46] If one turns to the simplest positive feedback loop ($\overset{+}{\alpha}$) one sees that it amounts to pure autocatalysis. If α is necessary and sufficient for its own synthesis, either it *was* already present and it will remain present indefinitely, or it was initially absent and it will remain absent indefinitely. Thus, this simple structure already accounts for multiple steady states, but it has much of a vicious circle; one understands that such a structure may have two stable states but one does not see how the choice between these two states could be operated. Slightly more complex structures account both for the multistationarity and for an understanding of the choice. For instance, a loop made of two antagonist elements ($\alpha \overset{\longrightarrow}{\longleftarrow} \beta$) has two stable states, one in which α (but not β) is present and steadily produced, and another in which β (but not α) is present and produced. Which state will be reached and stabilized from an initial state in which both products would be absent depends on the relative rates of appearance of the two products and of the levels of their thresholds of efficiency. In the logical description, all simple positive loops have two stable steady states (for specific cases: Refs. 1 and 15; for a formal demonstration: Ref. 16; any specific case can be treated according to the method described in Section II). Continuous analysis also predicts, for proper values of the parameters, two stable steady states that are attractors (and an unstable steady state).

To summarize this section: simple feedback loops behave differently depending on the parity of their number of negative interactions. Those with an *odd* number of negative interactions (negative loops) tend to stabilize the level of their elements around a submaximal value, usually by sustained oscillations around this value. The simple loops with an *even* number of negative interactions (positive loops) may exist in either of two stable states, in which an individual element is produced either at a very low level or near the maximal possible level. It is seen that the two types of simple feedback loops provide, respectively, for sustained oscillations and for multiple steady states, and, from the biological viewpoint, for the homeostatic and epigenetic type of regulation, respectively.

But real systems are usually not simple feedback loops. In a virus such as bacteriophage λ the decision to kill the infected bacterial cell or to establish a symbiotic association with it depends on complex interactions involving a number of interconnected feedback loops. Such systems (and even simpler ones) would need a formal description in view of their complexity; but as a matter of fact this complexity is such that the classical methods are much too heavy. This was a reason for trying a *logical* description, that is, a description using variables and functions which can take only a limited number of values—typically two (1 and 0).

II. A FULLY ASYNCHRONOUS LOGICAL DESCRIPTION AND TREATMENT OF SYSTEMS COMPRISING FEEDBACK LOOPS

A. The Involvement of Time in the Logical Description

The idea of treating complex systems of interacting elements in Boolean (= logical) terms is far from new: see, for instance, in the particular case of biological systems Refs. 17–19 and especially Refs. 15, 20, and 21.

The major problem is: How can one introduce time in the logical description of systems in a way that is at the same time coherent, realistic, and convenient? Time has usually been introduced in logical formalisms by giving the logical values of the variables "at time $t + 1$" as functions of their values "at time t." In practice, one tabulates the values at time $t + 1$ for each of the 2^n possible combinations of values of the n variables at time t; it is convenient to speak, for a network of n variables, in terms of an n vector whose value $\mathbf{x}_{t+1}$ at time $t + 1$ is given as a function of its value $\mathbf{x}_t$ at time t. This version is a synchronous one; where $\mathbf{x}_{t+1}$ and $\mathbf{x}_t$ differ by the values of more than one variable, these values are supposed to change in a synchronous way at time $t + 1$. The synchronous treatment is very easy but extremely unrealistic. One of its most negative aspects is that for each logical state there cannot be more than one next state: in this perspective, a system cannot diverge depending on the values of parameters or delays. Using logical equations of the type $[x_i(t + \tau_i) = f_i(x_1(t), x_2(t), \ldots, x_n(t))]$ is *not* a fully asynchronous treatment, if only because in this formalism the time delays to transit from $x_i = 0$ to $x_i = 1$ and from $x_i = 1$ to $x_i = 0$ are necessarily identical. As for the idea that one can bypass the difficulty by using such small τ_i that only one variable can switch within the interval, it is fallacious. Among the efforts to render practicable a *really* asynchronous treatment, I would like to mention especially the use of differential equations comprising a Boolean function[22,23], subsequently denoted as "PL" (piecewise linear ordinary differential) equations.[24] The authors combine this quantitative description with a logical analysis of the Boolean moiety of their PL equations, using "state transition diagrams" mapped on N-cubes[25] a development of the "toroid maps" of Glass and Kaufman.[15] The present paper deals with a *purely logical, yet fully asynchronous method* first described in Thomas[1] and Thomas and Van Ham.[2] This method is described in a more mathematical language in Milgram.[29]

B. The Logical Variables and Functions Used

Unlike other authors,* we choose from the beginning to characterize each element i of a system both by a logical *variable* α_i ("internal" variable) associated with its *concentration* (or, more generally, its level) and by a logical *function* a_i associated with its *rate of production* (or, more generally, its flux). Variable $\alpha = 1$ if the concentration exceeds a functional threshold, $\alpha = 0$ if not.† Function $a = 1$ if the rate of production

---

* The very nature of the Boolean variables used has varied considerably. For instance, in his initial papers Glass[15,25] associated with each element x_i of the system a Boolean variable whose value is 1 where $dx_i/dt > 0$ and 0 where $dx_i/dt < 0$. Subsequently, instead of using Boolean variables whose value was determined by the value of the time derivative of the corresponding continuous variable, Glass[22] shifted to Boolean variables $\bar{x}_i$ associated with the concentrations of the elements and such that $\bar{x}_i = 1$ if $x_i > \theta_i$ and $\bar{x}_i = 0$ if $x_i < \theta_i$ (like our variables α_i).

† May I insist on the point that $\alpha = 0$ does not mean that the concentration is nil, but that it is below the threshold?

(or activity) of the element is significant, $a = 0$ if not; for instance, in genetics $a = 1$ where a gene is "on," $a = 0$ where it is "off."

There is a well-defined time relation between the values of a function a_i and its associated variable α_i. Let us examine this relation, taking as an example a gene whose product is α. Clearly, in a steady state the logical values of a and α are the same: if a gene has been off ($a = 0$) for a long time, its product (which has a limited life span) will be "absent" ($\alpha = 0$); if the gene has been on ($a = 1$) for a sufficient time, the product will be present ($\alpha = 1$). If a change in the state of the variables results in a change of the value of a, α will adopt the new value of a, but only after a delay. In the meantime, the values of a and α will "disagree"* until α has adopted the new value of a. During this period, variable α is subject to an order to switch from its present value to the complementary value. For instance, when a gene is switched on, the product will appear, but not until some minutes (necessary for the synthesis and accumulation of the product), and when the gene is switched off its product will disappear, but only after a delay which depends on its life span, diffusibility, etc. Thus, we associate with each couple i (function a_i, variable α_i) *two* delays: one for the appearance and one for the disappearance of the product. Note that, formally, the relationship between a function a_i and its associated variable α_i is the same as the relationship between a logical function Y_i and its memorization variable y_i, as used in the field of sequential automata (Huffman[26]; Florine[27], whose work exerted a crucial influence on the development of our conceptions). Our finalities were, at least initially, very different. Conceiving a logical machine involves the *synthesis* of a logical structure which will behave according to well-defined desiderata. One step of this process consists of introducing internal functions Y_i and their memorization variables y_i, an artifice that permits treating sequential problems formally as if they were combinational ones. In this perspective, when the value of two or more memorization variables "disagree" with the values of the corresponding functions, one often reasons that they "should" shift simultaneously, but that in practice, due to small, irreproducible differences in the memory organs of the machine, one of the variables will switch first. Such a race is called a "critical race" if it leads the system to different final states depending on which variable switches first; critical races must of course be avoided by all means in logical machines. In this technology, an asynchronous treatment is needed, not so much per se, but mostly as a way to find networks that will behave in the specified way even if two (or more) variables due to

---

* Milgram[29] ably describes the situation as "*accord entre flux et mémoire*" when $a_i = \alpha_i$ and "*désaccord*" when the value of a_i has changed but α_i has not yet followed.

switch simultaneously do not. Thus, the formalism of Huffman and of Florine does permit an asynchronous treatment but only more recently was this possibility fully developed (see Van Ham[28]).

Since, in contrast, we wanted to simulate existing systems (more precisely, to analyze the dynamical possibilities of models), our functions, variables, and time delays *had* to have a concrete meaning. In this perspective, the occurrence of critical races is no more an unpleasant complication, but should play in the description of the systems an essential role, responsible for the multiplicity of the possible pathways. The idea of affecting each element of the system with a function describing its flux and a variable describing it concentration, and treating them, respectively, as a function and its memorization variable[1,2] was extremely helpful because it results in concrete functions, variables, and time delays.

An important feature of our method is the simplifying assumption that when, starting from a situation in which $a_i = \alpha_i$, one changes the value of a_i, the value of α_i will change after a characteristic delay *unless a counterorder* (an additional change of the value of a_i) *is given before the delay has elapsed*. If such a counterorder has taken place before the order has been executed, we reason as if the order had been merely cancelled.*[2,14,30] As all simplifying assumptions, this rule may introduce distortions in particular cases; however, the gain in simplicity by far compensates these occasional distortions.

C.   Logical Equations

Our logical equations usually describe systems by relating the logical values of the functions a_i (rates of production) with the values of the internal variables α_i (concentrations)† and of input variables such as temperature T:

$$a_i = \phi_i(\alpha_1, \alpha_2, \ldots \alpha_i, \ldots, \alpha_n; T, \ldots)$$

in which ϕ_i is a Boolean function. In a vectorial notation,

$$\mathbf{a} = \phi(\alpha; T)$$

If one compares these equations to the logical equations frequently used which relate the situation at time $t + 1$ to the situation at time t, one may ask where time is hidden in our equations. In fact, it is present in an

---

* In other words, we use "inertial" delays.[28]

† We do not exclude self-input (a_i is a function of the internal variables, possibly including α_i). In our formalism, the consideration of self-input, which considerably increases the possibilities, does not generate excessive complications.

implicit way (just as in the differential equations used in chemical kinetics) because one relates *rates* of production to concentrations. Time is, in addition, present in an *explicit* way as time delays which relate changes in the logical value of each concentration α_i with previous changes in the value of the corresponding rate of synthesis a_i. Here is an example, taken from Leclercq and Thomas,[31] of a logical scheme and the corresponding logical equations. The logical scheme

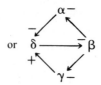

represents a system comprising two conjugated feedback loops, a negative one and a positive one. For instance, $\beta \rightarrow {}^-\alpha$ means that the condition for the synthesis of α is the absence of β, and

means that δ is synthesized iff α is absent or (inclusive) γ is present. The corresponding set of logical equations is

$$a = \overline{\beta}$$
$$b = \overline{\delta}$$
$$c = \overline{\beta}$$
$$d = \overline{\alpha} + \gamma$$

in which $\overline{\beta}$ means (not β) and $+$ is the logical sum (inclusive or). These equations are so simple that, at first view, one would hardly expect them to generate any complex dynamical behavior. This is an illusion, however, as we see below.

D. State Tables

From the logical equations that describe a system, we derive a state table that gives the values of the functions a_i for each of the 2^n combinations of values of the n variables α_i (Thomas[1,14,30]; Thomas and Van Ham[2]). In spite of a superficial similitude, these tables deeply differ from the classical ones because instead of tabulating x_{t+1} as a function of x_t we tabulate the logical values of the rates of production (fluxes) for each combination of values of the concentrations; thus, the Boolean "vector" in the right column is usually *not* the next state for the Boolean vector at the same line in the left column (see next section). Table I is the state table corresponding to the logical scheme and equations given above.

TABLE I.   State Table[a]

$\alpha\beta\gamma\delta$	abcd
$\bar{0}\bar{0}\bar{0}\bar{0}$	1111
$\bar{0}\bar{0}\bar{0}1$	1011
$\bar{0}\bar{0}11$	1011
$\bar{0}\bar{0}\bar{1}\bar{0}$	1111
$01\bar{1}\bar{0}$	0101
$0\bar{1}\bar{1}1$	0001
$0\bar{1}01$	0001
$010\bar{0}$	0101
$\bar{1}100$	0100
$\bar{1}1\bar{0}\bar{1}$	0000
$\bar{1}\bar{1}\bar{1}1$	0001
$\bar{1}1\bar{1}\bar{0}$	0101
$10\bar{1}\bar{0}$	1111
$\boxed{1011}$	1011
$100\bar{1}$	1010
$1\bar{0}\bar{0}\bar{0}$	1110

[a] For each of the 2^4 combinations of values of the 4 variables, this state table gives the values of the associated functions, as provided by the logical equations: $a = \bar{\beta}$; $b = \bar{\delta}$; $c = \bar{\beta}$; $d = \bar{\alpha} + \gamma$.

Where vectors $\boldsymbol{\alpha}$ and $\mathbf{a}$ are the same, in other words, where the value of each function "agrees" with the value of the corresponding variable, one deals with a *stable state* since no variable is requested to change its value. For instance, in Table I, for $\alpha\beta\gamma\delta = 1011$ one has also *abcd* = 1011; this is a stable state and we write $\boxed{1011}$. For the other states, one or more function "disagrees" with the corresponding variable, whose value is thus requested to change. We draw attention to this situation by a dash over the values of the variables that disagree with the corresponding functions. For instance, where $\alpha\beta\gamma\delta \mid abcd$ = 1110/0101 we write $\bar{1}1\bar{1}\bar{0}$. This compact notation* displays the fact that variables α and γ are requested to switch from 1 to 0 and variable δ from 0 to 1.

* The notation $\bar{1}1\bar{1}\bar{0}$/0101 (as used in Table I) is redundant but convenient for tables.

E.   Graph of the Sequences of States

In a synchronous system, the state following $\bar{1}\bar{1}\bar{1}\bar{0}$ would be 0101, with a synchronous change of the values of three variables. In addition to being unrealistic, this attitude prevents any choice between two or more possibilities: each state cannot have more than one possible next state. We reason rather[1] that, usually, *either* of the orders will be executed, thus leading in the present case to either of three possible next states:

$$\bar{1}\bar{1}\bar{1}\bar{0} \underset{\delta}{\overset{\bar{\alpha}}{\underset{\bar{\gamma}}{\rightleftarrows}}}
\begin{array}{l} 01\bar{1}\bar{0} \\ \bar{1}100 \\ \bar{1}\bar{1}\bar{1}\bar{1} \end{array}$$

where the dashes in the right members are taken from the state table. (Usually, we do not consider *explicity* the possibility that two or more commutations take place in exact simultaneity. Should this happen, the situation would be taken into account automatically in the simulations, however).

Consider transition $\bar{1}\bar{1}\bar{1}\bar{0} \xrightarrow{\bar{\gamma}} \bar{1}100$, in which variable γ switches from 1 to 0. In the resultant state $\bar{1}100$, the order $\bar{\alpha}$ is still present but the order δ has disappeared; and in accordance with the rule previously mentioned we reason as if this order to synthesize δ had been merely cancelled, because a counterorder (involved in the transition $\bar{1}\bar{1}\bar{1}\bar{0} \xrightarrow{\bar{\gamma}} \bar{1}100$) has taken place before the order has been executed.

In this way, one can derive graphs of the sequences of states. This type of derivation[1,2] is basically similar to the use of "toroid maps" independently proposed by Glass and Kauffman[12] and later generalized as "state transition" diagrams mapped on N-cubes.[25] We have occasionally used[2,30] a mapping on cubes, with the difference that we do not exclude self input, and, consequently, can have ridges with 0, 1, or 2 arrows rather than one and only one. However, we usually prefer to use open graphs in which a same Boolean state may occur more than once, because different occurrences of a same Boolean state may correspond to non-identical situations. For instance, in the sequence

$$\bar{0}\bar{0}\bar{0}\bar{0} \xrightarrow{\alpha} 1\bar{0}\bar{0}\bar{0} \underset{1\ 1}{\xrightarrow{\beta}} \bar{1}100 \xrightarrow{\bar{\alpha}} 010\bar{0} \xrightarrow{\delta} 0\bar{1}01 \xrightarrow{\bar{\beta}}$$

$$\bar{0}0\bar{0}1 \xrightarrow{\alpha} 10\bar{0}\bar{1} \underset{1}{\xrightarrow{\bar{\delta}}} 1\bar{0}\bar{0}0 \underset{2}{\xrightarrow{\beta}} \bar{1}100 \rightarrow \ldots$$

the binary state $1\bar{0}\bar{0}0$ occurs twice. These states are *not* equivalent since in the first case both orders β and γ were already present in the preceding state $(\bar{0}\bar{0}\bar{0}\bar{0})$ whereas in the second case order β was not yet present in the preceding state $(10\bar{0}\bar{1})$ but the order γ was already present two steps ahead (state $\bar{0}0\bar{0}1$). This can be expressed by subscripts[31] that indicate whether (and for how many steps) an order was already present in previous states; thus, in the first case we write $1\bar{0}00$ and in the second case $_{11}$

$1\bar{0}00$. Two states with different subscripts *are* different; two states with $_{2}$

the same index may be identical (as, for instance, in the two occurrences of the Boolean state $\bar{1}100$, which has *no* subscript in either case; in both cases, the delay $t_{\bar{\alpha}}$ begins to run exactly at the onset of state $\bar{1}100$).

F. Conditions of the Various Sequences of States. Logical Stability Analysis

Which pathway will be effectively followed depends on the initial state and on the relative values of the time delays. But what interests us is to determine in each case exactly in which way the Boolean trajectory depends on the delays, which delays are relevant, which are not, etc. Simple cases have been treated in Thomas[1] and Thomas and Van Ham.[2] Most of the procedure has been automatized.[32] Let us come back to our example. In a situation like

$$\rightarrow 0\bar{1}01 \xrightarrow{\bar{\beta}} \bar{0}0\bar{0}1 \begin{array}{c} \xrightarrow{\alpha} 10\bar{0}\bar{1}_{1} \begin{array}{c} \xrightarrow{\gamma} \boxed{101\bar{1}} \\ \xrightarrow{\bar{\delta}} 1\bar{0}\bar{0}0_{2} \end{array} \\ \xrightarrow{\gamma} \bar{0}011_{1} \xrightarrow{\alpha} \boxed{101\bar{1}} \end{array}$$

state $\bar{0}0\bar{0}1$ will be followed by $\xrightarrow{\alpha}_{1} 10\bar{0}\bar{1}$ or by $\xrightarrow{\gamma}_{1} \bar{0}011$ depending simply on whether the time delay t_{α} is shorter or longer than t_{γ}. From $10\bar{0}\bar{1}_{1}$, the system may proceed to $\xrightarrow{\gamma} \boxed{101\bar{1}}$ or to $\xrightarrow{\bar{\delta}} 1\bar{0}\bar{0}0_{2}$. In this case, the situation is slightly more complicated because the order to synthesize γ was already present during the preceding period whereas the order $\bar{\delta}$ has been given only at the onset of state $10\bar{0}\bar{1}_{1}$; thus, the decision depends on whether $t_{\gamma} < t_{\alpha} + t_{\bar{\delta}}$ or vice versa (here, the sign $+$ is the arithmetic, not the logical, sum !). With each fork in the graph of the sequences of states, one can associate inequalities (one, three, six, . . . depending on whether two, three, four, . . . functions "disagree" with their memorization variable) between the time delays or their entire linear combina-

tions. Thus, with any sequence of states one can associate a set of inequalities that give the conditions necessary and sufficient for this sequence
to be followed.

In particular, when one finds a state which has already been found
upstream (with the same subscript if any), one may determine in which
conditions this trajectory can be followed as a cycle.

Note that we do *not* consider as cycles situations like the following:

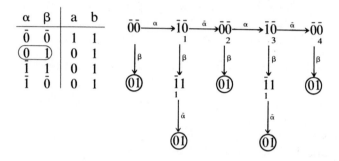

in which the successive occurrences of $\bar{0}\bar{0}$ (or of $\bar{1}\bar{0}$) are different, as shown
by the increasing value of their subscript. One easily sees that there can
be for some time an oscillation $\bar{0}\bar{0} \underset{\alpha}{\overset{\alpha}{\rightleftharpoons}} \bar{1}\bar{0}$ if the time delay t_β is long
enough, but sooner or later one will fall into the stable state $\textcircled{0 1}$. This
is apparently the logical equivalent of a damped oscillation.

One often finds cyclic sequences for which conditions are a set of
inequalities. These are *stable cycles,* since in the space of the delays
(which has $2n$ dimensions if there are n variables), there is a $2n$-dimen-
sional manifold within which the conditions are fulfilled; if one stands in
that domain, one can change the value of one or more delay and yet remain
within the range of the constraints. In other words, these cycles are, within
certain limits, stable toward alterations of the time delays. There exist
even cycles which, once reached, persist whatever the values of the time
delays. This situation takes place each time one deals with a cycle for
which each state has only one possible next state (only one dash).

For other cyclic sequences, one finds among the conditions one (or
more) *equality* between combinations of delays. These are *unstable cycles*
because, starting from the conditions which would permit the system to
go through them, the slightest modification of the value of one of the
delays involved in the equality would induce the system to leave this
trajectory. Another way to characterize this situation is to remark that
in the $2n$-dimensional space of the delays there is only a $2n - 1$ manifold
(hypersurface) within which the conditions are fulfilled.

As examples of our stability analysis, let us examine first the simplest cyclic trajectory in the double-loop system already considered above:

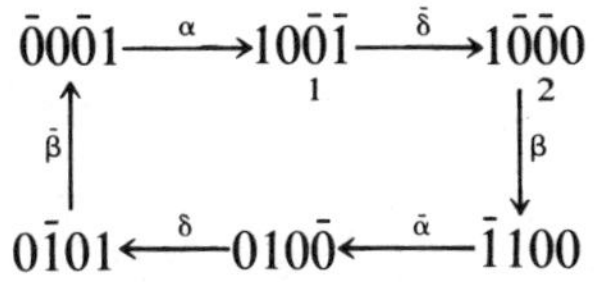

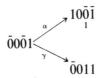

At the level of state $\bar{0}0\bar{0}1$, there is a fork:

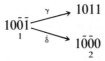

Whether the first or the second choice is adopted depends on whether or not $t_\alpha < t_\gamma$. We pose $m \equiv t_\alpha < t_\gamma$; $m = 1$ if the statement "$t_\alpha < t_\gamma$" is true, $m = 0$ if not. At the level of state $10\bar{0}\bar{1}$ there is a second fork:

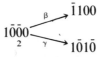

Whether the first or the second choice is adopted depends on whether or not $t_\gamma < t_\alpha + t_{\hat{\delta}}$. We pose $n \equiv t_\gamma < t_\alpha + t_{\hat{\delta}}$. At the level of state $1\bar{0}\bar{0}0$ there is a third and last fork:

Whether the first or the second choice is adopted depends on whether or not $t_\alpha + t_{\hat{\delta}} + t_\beta < t_\gamma$. We pose

$$p \equiv t_\alpha + t_{\hat{\delta}} + t_\beta < t_\gamma$$

Thus, the conditions to remain in this cyclic trajectory are $m\bar{n}p$. However, one can take advantage of the fact that variables m, n, and p are not independent to simplify the expression of the conditions of stability of the cycle:

$$m \equiv t_\alpha < t_\gamma \left.\vphantom{\begin{matrix}1\\1\end{matrix}}\right\} \quad \text{Thus, } \bar{n} \to m$$
$$n \equiv t_\gamma < t_\alpha + t_{\bar\delta}$$
$$p \equiv t_\alpha + t_{\bar\delta} + t_\beta < t_\gamma \left.\vphantom{1}\right\} \text{Thus, } p \to \bar{n}$$

Clearly, if $p \to \bar{n}$ and $\bar{n} \to m$, the conditions $m\bar{n}p$ reduce to p. Thus, the only condition to remain in this trajectory is $t_\alpha + t_{\bar\delta} + t_\beta < t_\gamma$. Consider now the cyclic sequence:

$$\underset{1}{1\bar{0}\bar{0}\bar{0}} \overset{\gamma}{\rule{1.5cm}{0.4pt}} \underset{1}{1\bar{0}1\bar{0}} \overset{\beta}{\rule{1.5cm}{0.4pt}} \underset{1}{\bar{1}1\bar{1}\bar{0}}$$

$$\underset{\bar\delta}{\uparrow} \hspace{6cm} \underset{\delta}{\downarrow}$$

$$\underset{1}{1\bar{0}\bar{0}\bar{1}} \underset{\bar\beta \quad 2\,1}{\rule{1.5cm}{0.4pt}} \bar{1}\bar{1}\bar{0}\bar{1} \underset{\bar\gamma \quad 1\ 1}{\rule{1.5cm}{0.4pt}} \bar{1}\bar{1}\bar{1}\bar{1}$$

First, notice that during part of this sequence there is an order $\bar{\alpha}$, but if the sequence is followed this order is never executed. The analysis is greatly simplified if one first assumes that the delay $t_{\bar\alpha}$ is such that variable α be never switched off, and only later details the conditions of this constraint.

Second, we see that, in contrast with all other cycles, *each* of the states of this sequence has at least one subscript, indicating that at least one function was already on before the onset of the state. Let us start arbitrarily from state $1\bar{0}\bar{0}\bar{0}$; the subscript shows that at the onset of this state the synthesis of γ was already on for some time, say ϕ ($\phi < t_\gamma$). The contraints that must be fulfilled in order that the cycle be followed are, at the level of state:

$1\bar{0}\bar{0}\bar{0}$	$(t_\gamma - \phi) < t_\beta$, or $t_\gamma < \phi + t_\beta$
$1\bar{0}1\bar{0}$	$\phi + t_\beta < t_\gamma + t_\delta$
$\bar{1}1\bar{0}$	$t_\gamma + t_\delta < \phi + t_\beta + t_{\bar\gamma}$
$\bar{1}\bar{1}\bar{1}$	$\phi + t_\beta + t_{\bar\gamma} < t_\gamma + t_\delta + t_{\bar\beta}$
$\bar{1}\bar{1}0\bar{1}$	$t_\gamma + t_\delta + t_{\bar\beta} < \phi + t_\beta + t_{\bar\gamma} + t_{\bar\delta}$
$1\bar{0}\bar{0}\bar{1}$	$\phi + t_\beta + t_{\bar\gamma} + t_{\bar\delta} < t_\gamma + t_\delta + t_{\bar\beta} + t_\gamma^{*}$
$1\bar{0}\bar{0}\bar{0}$ (2nd occurrence)	$t_\gamma + t_\delta + t_{\bar\beta} + t_\gamma < \phi + t_\beta + t_{\bar\gamma} + t_{\bar\delta} + t_\beta$
$1\bar{0}\bar{0}\bar{0}$ [$(k + 1)$th occurrence)]	$k(t_\gamma + t_\delta + t_{\bar\beta}) + t_\gamma < \phi + k(t_\beta + t_{\bar\gamma} + t_{\bar\delta}) + t_\beta$

---

* In this analysis, we assume that a given delay will take the same value for its successive occurrences.

Let us set $K = t_\gamma + t_\delta + t_{\bar\beta}$, $L = t_\beta + t_{\bar\gamma} + t_{\bar\delta}$ and $\Delta = K - L$. One can write that at the $(k + 1)$th occurrence of state $1\bar00\bar0$, the condition to

remain in the cycle is

$$kK + t_\gamma < \phi + kL + t_\beta$$

or

$$k\Delta + t_\gamma < \phi + t_\beta \tag{1}$$

and at the $(k + 1)$th occurrence of state $1\bar010\bar0$ the condition is

$$\phi + kL + t_\beta < kK + t_\gamma + t_\delta$$

or

$$\phi + t_\beta < k\Delta + t_\gamma + t_\delta \tag{2}$$

(in these inequalities, all the terms except Δ are positive). If Δ is positive, one can always find a value of k high enough that inequality (1) is not fulfilled, and if Δ is negative one can always find a value of k high enough that inequality (2) is not fulfilled. Thus, in order to remain indefinitely in this cycle, one of the conditions is $\Delta = 0$, in other words:

$$t_\gamma + t_\delta + t_{\bar\beta} = t_\beta + t_{\bar\gamma} + t_{\bar\delta} \tag{3}$$

Granted the other conditions,* the system will remain in the cycle when the two sums of delays are exactly equal but the smallest change of one of the delays involved in (3) will eventually result in the system leaving

---

* The other conditions are

$$t_\gamma < \phi + t_\beta < t_\gamma + t_\delta < \phi + t_\beta + t_{\bar\gamma} < t_\gamma + t_\delta + t_{\bar\beta};$$

$t_{\bar\gamma} < t_\alpha$ and (for any value of $k \in Z_0$)

$$t_\gamma + t_\delta + t_{\bar\beta} + kK < t_\alpha + t_{\bar\delta} + t_\beta + t_{\bar\alpha} + kL \tag{4}$$

but when condition (3) is fulfilled $K = L$ and (4) simplifies into

$$t_\gamma + t_\delta + t_{\bar\beta} < t_\alpha + t_{\bar\delta} + t_\beta + t_{\bar\alpha}$$

the cycle; as already mentioned above, when the conditions include one (or more) equality, one deals with an *unstable cycle*.

G.   The Final States

The graph of the sequences of states may lead to one or more final states (attractors), each of which may be a stable state (already detected at the level of the state tables), a cyclic attractor, or perhaps a non-periodic attractor (e.g., a chaotic situation).

As far as we can tell, a necessary (but not sufficient) condition for multiple attractors is that the logical structure comprise at least one positive loop, and a necessary (but not sufficient) condition for a stable periodic behavior is the presence of at least one negative loop in the logical structure.* For instance, a simple positive loop can provide for two attractors (in this case, two stable states). In order to have *both* multiple attractors and cyclic behaviors, one needs at least two loops, a positive and a negative one; the system taken as an example in this chapter provides for two attractors, a stable state and a cyclic attractor. *Note that except for very simple systems a cyclic attractor may display various sequences of boolean states depending on the values of the time delays.* This point was already mentioned, but in a somewhat ambiguous way in a former paper,[14] in which I state that in the case of a system composed of two conjugated negative loops "there may be, instead of a single, stable cycle, two or more distinct cycles." In that case (two conjugated negative loops) as well as in the example treated throughout the present paper (two conjugated loops: a positive and a negative one) there may indeed be several stable cyclic sequences of states (stable cycles). However, these cycles have states in common and they are thus mere variants of a same cyclic attractor† (see Leclercq and Thomas[31]), just in the same way as, in the continuous description, a limit cycle has various shapes depending on the values of the parameters.

Taking again as an example the system defined in Section II.C, here follow the seven *simple* stable cycles of this system, that is, those stable cycles in which no Boolean state occurs more than once in a period:

C1   (1) $\alpha\bar{\delta}\beta\bar{\alpha}\delta\bar{\beta}$
      1-11-10-14-4-5
      (octal representation
      of the Boolean states)

$$\bar{0}0\bar{0}1 \xrightarrow{\alpha} 10\bar{0}\bar{1} \xrightarrow{\bar{\delta}} 1\bar{0}\bar{0}0$$
$$\scriptstyle\bar{\beta}\uparrow \qquad\qquad {\scriptstyle 1} \qquad\qquad {\scriptstyle \downarrow\beta \atop 2}$$
$$0\bar{1}01 \xleftarrow{\delta} 010\bar{0} \xleftarrow{\bar{\alpha}} \bar{1}100$$

---

* A formal demonstration is proposed by Van Ham.[44]

† In contrast, we will see in Section III.D a more complex case in which there are really two distinct cyclic attractors; in that case there is no Boolean state common to the two attractors and for identical values of the delays the system may follow one or the other cycle depending on the initial state.

C2   (1) $\alpha\delta\gamma\beta\bar{\alpha}\bar{\gamma}\delta\bar{\beta}$
       1-11-10-12-16-6-4-5

$$\bar{0}001 \xrightarrow{\alpha} 10\bar{0}\bar{1} \xrightarrow{\bar{\delta}} 1\bar{0}00 \xrightarrow{\gamma} 1\bar{0}1\bar{0}$$
$$\quad\;\;\; {}_{1} \qquad\qquad {}_{2} \qquad\qquad {}_{1}$$
$$\uparrow\bar{\beta} \qquad\qquad\qquad\qquad\qquad\qquad \downarrow\beta$$
$$0\bar{1}01 \xleftarrow{\delta} 0100 \xleftarrow{\bar{\gamma}} 01\bar{1}\bar{0} \xleftarrow{\bar{\alpha}} \bar{1}1\bar{1}0$$
$$\qquad\qquad {}_{3} \qquad\quad {}_{1\,2} \qquad\qquad {}_{1}$$

C3   (1) $\alpha\delta\gamma\beta\bar{\alpha}\delta\bar{\gamma}\bar{\beta}$
       1-11-10-12-16-6-7-5

$$\bar{0}001 \xrightarrow{\alpha} 10\bar{0}\bar{1} \xrightarrow{\bar{\delta}} 1\bar{0}00 \xrightarrow{\gamma} 1\bar{0}1\bar{0}$$
$$\quad\;\;\; {}_{1} \qquad\qquad {}_{2} \qquad\qquad {}_{1}$$
$$\uparrow\bar{\beta} \qquad\qquad\qquad\qquad\qquad\qquad \downarrow\beta$$
$$0\bar{1}01 \xleftarrow{\bar{\gamma}} 0\bar{1}\bar{1}1 \xleftarrow{\delta} 01\bar{1}\bar{0} \xleftarrow{\bar{\alpha}} \bar{1}1\bar{1}0$$
$$\quad {}_{1} \qquad\quad {}_{2} \qquad\quad {}_{1\,2} \qquad\qquad {}_{1}$$

C4   (1) $\alpha\delta\gamma\beta\bar{\gamma}\bar{\alpha}\delta\bar{\beta}$
       1-11-10-12-16-14-4-5

$$\bar{0}001 \xrightarrow{\alpha} 10\bar{0}\bar{1} \xrightarrow{\bar{\delta}} 1\bar{0}00 \xrightarrow{\gamma} 1\bar{0}1\bar{0}$$
$$\quad\;\;\; {}_{1} \qquad\qquad {}_{2} \qquad\qquad {}_{1}$$
$$\uparrow\bar{\beta} \qquad\qquad\qquad\qquad\qquad\qquad \downarrow\beta$$
$$0\bar{1}01 \xleftarrow{\delta} 0100 \xleftarrow{\bar{\alpha}} \bar{1}100 \xleftarrow{\bar{\gamma}} \bar{1}1\bar{1}0$$
$$\qquad\qquad {}_{1} \qquad\qquad {}_{1} \qquad\qquad {}_{1}$$

C5   (1) $\alpha\delta\gamma\beta\delta\bar{\alpha}\bar{\gamma}\bar{\beta}$
       1-11-10-12-16-17-7-5

$$\bar{0}001 \xrightarrow{\alpha} 10\bar{0}\bar{1} \xrightarrow{\bar{\delta}} 1\bar{0}00 \xrightarrow{\gamma} 1\bar{0}1\bar{0}$$
$$\quad\;\;\; {}_{1} \qquad\qquad {}_{2} \qquad\qquad {}_{1}$$
$$\uparrow\bar{\beta} \qquad\qquad\qquad\qquad\qquad\qquad \downarrow\beta$$
$$0\bar{1}01 \xleftarrow{\bar{\gamma}} 0\bar{1}\bar{1}1 \xleftarrow{\bar{\alpha}} \bar{1}\bar{1}\bar{1}1 \xleftarrow{\delta} \bar{1}1\bar{1}0$$
$$\quad {}_{2} \qquad\quad {}_{1\,2} \qquad\quad {}_{1\;1} \qquad\qquad {}_{1}$$

C6   (1) $\alpha\delta\gamma\beta\delta\bar{\gamma}\bar{\alpha}\bar{\beta}$
       1-11-10-12-16-17-15-5

$$\bar{0}001 \xrightarrow{\alpha} 10\bar{0}\bar{1} \xrightarrow{\bar{\delta}} 1\bar{0}00 \xrightarrow{\gamma} 1\bar{0}1\bar{0}$$
$$\quad\;\;\; {}_{1} \qquad\qquad {}_{2} \qquad\qquad {}_{1}$$
$$\uparrow\bar{\beta} \qquad\qquad\qquad\qquad\qquad\qquad \downarrow\beta$$
$$0\bar{1}01 \xleftarrow{\bar{\alpha}} \bar{1}\bar{1}0\bar{1} \xleftarrow{\bar{\gamma}} \bar{1}\bar{1}\bar{1}1 \xleftarrow{\delta} \bar{1}1\bar{1}0$$
$$\quad {}_{2} \qquad\quad {}_{2\,1} \qquad\quad {}_{1\;1} \qquad\qquad {}_{1}$$

C7   (1) $\alpha\delta\gamma\beta\delta\bar{\gamma}\bar{\delta}\bar{\alpha}\delta\bar{\beta}$
       1-11-10-12-16-17-15-14-4-5

$$\bar{0}001 \xrightarrow{\alpha} 10\bar{0}\bar{1} \xrightarrow{\bar{\delta}} 1\bar{0}00 \xrightarrow{\gamma} 1\bar{0}1\bar{0} \xrightarrow{\beta} \bar{1}1\bar{1}0$$
$$\quad\;\;\; {}_{1} \qquad\qquad {}_{2} \qquad\qquad {}_{1} \qquad\qquad {}_{1}$$
$$\uparrow\bar{\beta} \qquad\qquad\qquad\qquad\qquad\qquad\qquad\qquad\qquad \downarrow\delta$$
$$0\bar{1}01 \xleftarrow{\delta} 0100 \xleftarrow{\bar{\alpha}} \bar{1}100 \xleftarrow{\bar{\delta}} \bar{1}\bar{1}0\bar{1} \xleftarrow{\bar{\gamma}} \bar{1}\bar{1}\bar{1}1$$
$$\qquad\qquad {}_{3} \qquad\qquad {}_{2\,1} \qquad\quad {}_{1\;1}$$

Note that all these trajectories have in common the sequence

$$\longrightarrow 0\bar{1}01 \xrightarrow{\bar{\beta}} \bar{0}0\bar{0}1 \xrightarrow{\alpha} 10\bar{0}\bar{1} \xrightarrow{\bar{\delta}} 1\bar{0}\bar{0}0 \longrightarrow$$
$$\qquad\qquad\qquad\qquad\qquad\quad {}_{1} \qquad\qquad {}_{2}$$

They differ from each other by the number and order of the other commutations of the variables. One of them (C_1) has only six states (there is no commutation of variable γ). Five $(C_2$ to $C_6)$ have eight states; each of the four variables commutes back and forth. One variant (C_7) has ten states; one of the variables (δ) commutes four times, and displays thus a double periodicity. Multiple periodicities will be found on a much larger scale below.

It was felt initially that unstable cycles have only an academic interest in the sense that they cannot be final states. However, I realized later that *compound* stable cycles can be derived from each simple stable cycle, by incorporation of 1, 2, 3, ... m, ... runs of the unstable cycle. For instance, from the simple stable cycle C_1^o, one can derive an unlimited number of compound cycles (C_1^1, C_1^2, ... C_1^m, ...):

$$[1\bar{0}1\bar{0}\overset{\beta}{\rule{1.2em}{0.4pt}}\bar{1}1\bar{1}\bar{0}\overset{\delta}{\rule{1.2em}{0.4pt}}\bar{1}\bar{1}\bar{1}1\overset{\bar{\gamma}}{\rule{1.2em}{0.4pt}}\bar{1}\bar{1}0\bar{1}\overset{\beta}{\rule{1.2em}{0.4pt}}10\bar{0}\bar{1}\overset{\bar{\delta}}{\rule{1.2em}{0.4pt}}1\bar{0}\bar{0}0]_m$$

$$1\bar{0}\bar{0}0\overset{\bar{\delta}}{\rule{1.2em}{0.4pt}}10\bar{0}\bar{1}\overset{\alpha}{\rule{1.2em}{0.4pt}}\bar{0}0\bar{0}\bar{1}\overset{\beta}{\rule{1.2em}{0.4pt}}\bar{0}\bar{1}01\overset{\delta}{\rule{1.2em}{0.4pt}}\bar{0}10\bar{0}\overset{\bar{\alpha}}{\rule{1.2em}{0.4pt}}\bar{1}1100$$

in which m can take any integer value including 0. The logical stability analysis shows that these compound cycles are stable ones. For each one, there is a $2n$-dimensional (eight in the present case) manifold in the $2n$-dimensional space of the delays; the "volume" occupied by these domains of stability is thinner and thinner as m increases. As m tends to ∞, one tends to the unstable cycle, which occupies a $(2n - 1)$-dimensional manifold in the $2n$-dimensional space of the delays. Thus, in the example chosen, the unique cyclic attractor is composed of seven families of variants, which allow for an unlimited variety of periodicities.

The simplest way to identify the compound stable cycles has consisted of wiring up our system on the logical simulator Delphine,[36] starting from an initial state within the unstable cycle* and time delays such that the unstable state would persist; then changing (one at a time) each of the relevant delays in either direction. When one changes a relevant delay by, say, 1%, the system goes about 100 times through the unstable cycle, then, depending on the case, it falls into the stable state or follows a compound stable cycle.

One can as well obtain these situations using the computer programs elaborated by Van Ham and Dehoeck,[32] which identify a trajectory as a cycle when they find a logical state already found with exactly the same quantitative situation of the delays.

---

* As mentioned previously, each state of the unstable cycle has one (or more) subscript, indicating that one of the time delays has been running already for some time. Thus, whatever the state chosen to begin with, the delay in question must appear for the first time with the value $(t - \phi)$ instead of t (see Section II.F).

III.   RELATIONS WITH OTHER DESCRIPTIONS, APPLICABILITY TO OTHER FIELDS, AND INDUCTIVE VERSUS DEDUCTIVE USE

A.   On the Relation Between the Discrete Description and the Continuous Description Using Differential Equations

The differential equations used for the description of regulatory systems are frequently of the type

$$\dot{x}_i = f_i(x_1, x_2, \ldots, x_n) - k_{-i}x_i$$

in which f_i is a suitable combination of sigmoid functions and $-k_{-i}x_i$ is a term of decay, assumed to be proportional to the concentration of the substance[33,34]. As pointed out by Walter[35] and by Glass and Kaufman,[12] for example, these functions tend rapidly to step functions as the sigmoid becomes steeper and steeper.

In order to establish the relation between this type of differential equation and our logical equations, one can rewrite the differential equations as

$$\dot{x}_i + k_{-i}x_i = f_i(x_1, x_2, \ldots, x_n)$$

and remark that the second member is homologous to the second member of our logical equations:

$$a_i = \phi_i(\alpha_1, \alpha_2, \ldots, \alpha_n)$$

This amounts to saying that our functions a_i are the logical equivalents *not* of the time derivatives $\dot{x}_i$, but of $\dot{x}_1 + k_{-i}x_i$, which represents the *gross* rate of synthesis of x_i (not corrected for the decay term $-k_{-i}x_i$). In other words, the leak of x is not described *explicitly* in the logical equation; its value is taken into account in the time delays.

With this in mind, a system of logical equations (below, left) can be translated into a homologous system of differential equations (below, right) as follows:

$$a = \bar{\beta} \qquad \dot{x}_1 = f_1^-(x_2) - k_{-1}x_1$$

$$b = \bar{\delta} \qquad \dot{x}_2 = f_2^-(x_4) - k_{-2}x_2$$

$$c = \bar{\beta} \qquad \dot{x}_3 = f_3^-(x_2) - k_{-3}x_3$$

$$d = \bar{\alpha} + \gamma \qquad \dot{x}_4 = f_4^-(x_1) + f_5^+(x_3) - k_{-4}x_4$$

in which $f_1^-(x_2)$ represents a decreasing sigmoid such as

$$\frac{k_1}{1 + (x_2/\theta_2)^n}$$

and $f_5^+(x_3)$ represents an increasing sigmoid such as (see, e.g., Ref. 15)

$$\frac{k_5(x_3/\theta_3)^n}{1 + (x_3/\theta_3)^n}$$

$\theta_1, \theta_2, \ldots$ are the threshold concentrations of $x_1, x_2, \ldots$, respectively. Note that where a same variable retroacts at different levels (e.g., x_2 which acts on $\dot{x}_1$ and on $\dot{x}_3$), either one should use the same threshold value (here, θ_2) in the various instances, or, if there are reasons to have distinct thresholds, one should associate more than one logical variable (one for each threshold) with the continuous variable. Note also that the logical expression $\overline{\alpha} + \gamma$ has been translated simply into $f^-(x_1) + f^+(x_3)$; similarly, $\alpha \cdot \gamma$ would have been translated into $f^-(x_1) \cdot f^+(x_3)$. This is because the arithmetic sum and product apply to sigmoid functions essentially in the same way as the logical sum and product (respectively) to step functions.

In the continuous formalism, the determination of the trajectories involves an integration, which can be made analytically only in very simple cases, but has to be made numerically in most practical situations; when this is the case, it may be difficult to have a general idea of the dynamical possibilities of the system. Much in the same way as in the continuous treatment, we can ascribe numerical values to the delays (this is the "numerical" approach) and use the logical simulator "Delphine" of Van Ham[36] or computer programs as in Van Ham and Dehoeck[32]. One can also proceed without ascribing any numerical value to the time delays. It is tempting for me to call this approach "analytical": although the derivation of the logical trajectories involves an algorithmic procedure it does not involve the use of numerical values; and similarly, the logical stability analysis does not imply the assignment of numerical values to the time delays.

A key of the efficiency of our method (and the reason why it can deal with complex systems) is that the set of trajectories consistent with a logical structure can always be obtained in the analytical mode:

1. For each state, one knows from the state table which next states are conceivable in the system.

2. Any sequence can be analyzed in order to see in which conditions (if any) it will be followed.

As remarked in Section II.E an essential feature of the asynchronous treatment is the fact that certain Boolean states have two or more possible next states. Note that this has no philosophical pretentions at the level of the problem of determinism: a Boolean state may have more than one possible next state depending on the time delays, essentially in the same way as, in the continuous description, different values of the parameters may result in different trajectories. But, in addition, it must be realized that a Boolean state is not punctual; it would correspond, in the continuous description, to a whole domain in the space of variables. More specifically, in a continuous description, one can cut the n-dimensional space of variables into 2^n boxes, each of which corresponds to a state of the Boolean description. In the continuous space of variables a state is entirely defined by its coordinates; in the Boolean description, it is defined in a broad way by a Boolean number and in more detail by data about the carryover of the time delays. Finally, the simplicity of the Boolean treatment permits to include in an especially easy way the possibility that the values of the time delays fluctuate with time (of course, this has more to do with the problem of determinism).

As regards the correspondance between the predictions of our Boolean description and of the continuous description, one does not expect them to fit entirely, in view of the different simplifying assumptions. An important paper by Glass and Kauffman[12] shows, however, that the transition from a sigmoid interaction to a Heaviside (which is all-or-none) interaction does not alter the qualitative dynamics of the system in the cases chosen. There are some problems. For instance, what is described in Boolean descriptions as a stable cycle may be in the continuous description a limit cycle or a stable steady state depending on the values of the parameters or delays. This problem was analyzed at depth by Richelle[11,37] (see also Ref. 38). Let us simply recall here that in living processes the occurrence of macromolecular syntheses practically forces one to take incompressible time lags into account, and once this is done the constraints for obtaining permanent oscillations are largely relaxed; another simple consideration is that where the logical analysis predicts a permanent oscillation and the continuous analysis a stable steady state, the former may be right at the level of the individual cells and the latter at the level of the population.

In the example used throughout this paper, the logical analysis predicts a choice between two attractors (a stable state and a stable cycle), depending on the initial state and values of the time delays (with the theoretical possibility of an unstable cycle between the two attractors). The translation into the continuous language is: "For proper values of the parameters there is a choice (depending on the initial state and values of the parameters) between two attractors; one of them is a stable state and the second is a limit cycle or a stable steady state depending on the values of the parameters; and between these two attractors there is an unstable situation which may be cyclic." The continuous system is already too complex to be treated analytically. However, starting from the logical analysis, it has been possible to find values of the parameters such that the continuous system has indeed the choice between a stable steady state and a limit cycle, depending on the initial state.[31] Finally, although the unstable cycle predicted by the logical analysis has not been reproduced so far by numerical integration of a continuous system, we have recently shown (unpublished) that in proper conditions the unstable steady state (which is repulsive for any trajectory outside the separatrix) is approached in an oscillatory way by trajectories close to the separatrix; moreover, the oscillatory approach fits qualitatively with the Boolean predictions. Note that in simple cases we have now approximate relations that provide a fair estimation of the time delays from the parameters and vice versa.

B. Numerical and Logical Iterations

At this point, it may be of interest to briefly review the status of various types of iterative methods, in the Boolean as well as in the continuous description.

There are iterative methods (e.g., Jacobi, Gauss-Seidel, Newton) whose purpose is simply to provide solutions for the *steady-state equations*; others (e.g., Euler and its improved versions) aim to give *trajectories*. Cycling will be felt as a disagreeable iteration artifact in the first case, as an indication of a probably cyclic trajectory in the second case. The relation between the behavior in a simple iteration method (e.g., Jacobi) and the real trajectory is interesting, if not simple. Consider, for instance, a simple negative loop comprising three inhibitory elements:

$$x = f_1^-(z) - k_{-1}x$$
$$y = f_2^-(x) - k_{-2}y \qquad\qquad (5)$$
$$z = f_3^-(y) - k_{-3}z$$

This system admits a single steady state which may be stable or unstable depending on the values of the parameters; where the steady state is unstable, the system tends to a limit cycle.

The steady state equations are

$$x = \frac{f_1^-(z)}{k_{-1}}$$

$$y = \frac{f_2^-(x)}{k_{-2}} \tag{6}$$

$$z = \frac{f_3^-(y)}{k_{-3}}$$

A Jacobi or Gauss–Seidel* iteration on (6) will provide us with the co-ordinates of the steady state, or it will cycle indefinitely, depending on the slope of the functions. On the other hand, determining the trajectory by numerical integration of equation (5) will lead to a stable steady state or to a limit cycle depending on the slope of the functions. There is thus an obvious formal similarity between the two situations. However, the steepness corresponding to the transition from a punctual to a cyclic attractor is much smaller in the first case (in which the cyclic attractor is an iteration artifact) as in the second case (in which the cyclic attractor is close to the real trajectory).

The situation is rather similar if one applies iterative methods to the Boolean description. As noticed by, for example, Robert[39] and by Goles,[40] Boolean iterations "in parallel" and "in series" correspond, respectively, to the Jacobi and Gauss–Seidel iterations used in the quantitative description. In the first case (Jacobi), from an initial Boolean state ($\alpha_0\beta_0\gamma_0$, . . .) one computes† the values of the functions a, b, c, . . . which are reintroduced, respectively, as $\alpha_1\beta_1\gamma_1$, . . . , and so on; in the second case (Gauss–Seidel), the new value of each variable is reintroduced in a defined (but arbitrary) order.

Just in the same way as in the continuous description, these methods may lead to stable states, and where they lead instead to cyclic situations these may be mere iteration artifacts or not, depending on the case. For

---

* In the Jacobi iteration, one introduces arbitrary initial values (say x_1, y_1, z_1) of x, y, z in the right-hand sides of equations (6); this provides (left side) a new set of values (x_2, y_2, z_2) which are reintroduced in the right sides and so on. In the Gauss–Seidel iteration the new value of a variable is reintroduced in the next equation as soon as it has been computed; from x_1, y_1, z_1 one calculates x_2, from x_2, y_1, z_1, one calculates y_2, and so on.

† One can easily make pocket calculator programs which are the same (except of course for the functions themselves) for the iteration of logical and of continuous functions.

instance, in the corresponding logical system:

$$a = \bar{\gamma}$$
$$b = \bar{\alpha}$$
$$c = \bar{\beta}$$

$\alpha\beta\gamma$	abc
$\bar{0}\bar{0}\bar{0}$	111
$0\bar{0}1$	011
$01\bar{1}$	010
$\bar{0}10$	110
$1\bar{1}0$	100
$\bar{1}\bar{1}\bar{1}$	000
$\bar{1}01$	001
$10\bar{0}$	101

a Jacobi (parallel) iteration leads to $\bar{0}\bar{0}\bar{0} \Leftrightarrow \bar{1}\bar{1}\bar{1}$, which is an iteration artifact or to

$$0\bar{0}1 \rightarrow 01\bar{1} \rightarrow \bar{0}10 \rightarrow 1\bar{1}0 \rightarrow 10\bar{0} \rightarrow \bar{1}01,$$

the logical representation of the above-mentioned limit cycle.

Note that the synchronous treatment of a Boolean system is in fact a Jacobi (parallel) iteration. Our treatment may also be considered a kind of iteration, but it is neither a Jacobi iteration (in which all commutations are synchronous) nor a Gauss–Seidel iteration (in which the commutations take place one at a time but in a predetermined, arbitrary, order). We consider all the successions of states implicitly contained in the state table; which one is followed depends on the values of the delays.

C. To What Extent Can This Methodology Be Applied to Chemical Systems?

As already mentioned, the type of analysis described here is convenient to treat systems whose elements interact with each other in a *catalytic* way, that is, they influence the rate of synthesis of each other without transforming into each other. How could one describe by logical equations situations in which elements of the system *transform* into each other, as in usual chemical reactions?

One way to include these situations would be to realize that a substance x can exert a positive effect on the synthesis of substance y in two ways: x may exert a catalytic effect (if a chemical reaction is symbolized $\Rightarrow$, we would write $\overset{x}{\Rightarrow} y$); or x may transform into y (in the same symbolism, we would write $x \Rightarrow y$). How could we represent these two situations in terms of logical schemes? In the first case there is no problem: we write that x exerts a positive control on the synthesis of y ($x \longrightarrow^{+} y$). In the second case, one might argue (e.g., Thomas[30,41]) that x exerts a positive effect on the synthesis of y but this time *at its own* expense; thus, it exerts a positive action on the synthesis of y and at the same time a negative effect on its own net synthesis:

$$x \xrightarrow{\ +\ } y$$

However, a different view emerged from a discussion with Jeener. Let us consider the system:

$$\xRightarrow{k_1} x \xRightarrow{k_2} y \xRightarrow{k_3} z$$

in which x transforms into y and y into z according to a linear kinetics, and z exerts a negative feedback on the rate of synthesis of x:

$$\dot{x} = \frac{k_1}{1 + (z/\theta_z)^n} - k_2 x - k_{-1} x$$

$$\dot{y} = k_2 x - k_3 y - k_{-2} y \qquad (7)$$

$$\dot{z} = k_3 y - k_{-3} z$$

If one poses $k_2 x + k_{-1} x = K_1 x$ and $k_3 y + k_{-2} y = K_2 y$, the equations become formally identical with those representing the simple feedback loop

(in its variant in which x exerts a linear positive control on the synthesis of y, y a linear positive control or the synthesis of z, and z a sigmoid negative control on the synthesis of x). In both cases the system will have a single steady state; for proper values of n and of the k_i and k_{-i}, this steady state will be unstable and the final state will be a limit cycle. It seems thus reasonable to represent both systems by the same set of logical equations. This amounts to describing the conversion of x into y simply as a positive control of x on the synthesis of y, without mentioning explicitly in the logical equations that this synthesis of y takes place at the expense of x. But this loss of x is taken into account in the determination of the values of the time delays related to x: the kinetic constant k_{-1} of the continuous description is replaced by $K = k_2 + k_{-1}$ and the time delays related to x are modified in accordance.

As suggested by Nicolis, this attitude might be extended even to cases in which x is converted into y according to a nonlinear kinetics.

However, when *another* substance, say u, exerts a positive control on the transformation of x into y, I would write explicitly not only that x and u exert a positive control on y but also that u exerts a negative control on x. For instance, the model of synthesis of cyclic AMP developed by Goldbeter, which we already considered in simplified form (Thomas,[30,41])

would now be described in logical terms:

in which negative actions of β and γ on α are explicitly mentioned but the negative action of α on itself is not indicated explicitly:

$$a = \overline{\beta}\overline{\gamma}$$
$$b = \alpha\gamma$$
$$c = \alpha\beta$$

At any rate, a possible application of our formalism to chemical systems involving not only catalytic interations but also interconversions of elements will require much more work, including consideration of how to take care of stoichiometry.

D. Synthetic (Inductive) Mode

So far, we started from a logical structure and asked which final states (stable or periodic) are consistent with this structure, by which sequences of states they can be reached and which conditions determine the pathway that will be followed. In opposition with this analytic, deductive approach, one can start from an observed (or desired) behavior and try to proceed as rationally as possible toward logical structures that would permit, or impose, this behavior.*

In other words, one may ask to what extent one can proceed rationally to the induction of a model from the observed behavior, a process that usually requires more intuition than reasoning. This endeavor may seem quite pretentious at first view. However, its ambitions are somewhat restricted where one can reasonably assume that the essential elements of the system have been identified; what then are the simplest interactions between these elements which would account for the observed behavior (or impose it)? Obviously, even in this case the solutions are multifold. A simple case can be found in Thomas.[41] Examples taken from the work of Friesen and Stent[42] on the nervous system of the leech had been treated previously.[30] More complex examples (12 variables) in the same field have been treated since.

Here is another example: we want to find a four-element logical structure such that, depending on the initial state, the system would have (whatever the time delays) to follow either of two cyclic trajectories running in opposite directions, more specifically:

$$
\begin{array}{lll}
0001 \xrightarrow{\gamma} 0011 \xrightarrow{\bar{\delta}} 0010 & \quad & 1001 \xleftarrow{\bar{\gamma}} 1011 \xleftarrow{\delta} 1010 \\
\bar{\beta}\uparrow \qquad\qquad\qquad \downarrow\beta & \text{or} & \beta\downarrow \qquad\qquad\qquad \uparrow\bar{\beta} \\
0101 \xleftarrow{\delta} 0100 \xleftarrow{\bar{\gamma}} 0110 & \quad & 1101 \xrightarrow{\bar{\delta}} 1100 \xrightarrow{\gamma} 1110
\end{array}
$$

In this particular case, each state in a cycle has only one possible next state. For instance, 0001 is necessarily followed by 0011; to ensure that result we write 00$\bar{0}$1, which means that in state 0001 the *only* order is γ. One can reason in the same way for each of the 12 states in the cycles. This allows us to write a partial state table (Table IIa). Note that in this table there is no specification ($----$) for the value of the functions in states 0000, 0111, 1111, and 1000, except that we do *not* want them to be *stable* states.† In other words, each of the dashes in Table IIa could be replaced by a 1 or a 0 (except for the combinations which would create stable states); as there are 16 dashes, there is thus a total of almost 2^{16}, that is about 6×10^4 possibilities!

---

* We mentioned in (Section II.A) the influence exerted on this work by people who conceive logical machines (notably J. Florine). It was noted that our initial purposes were widely different (theirs essentially synthetic, our essentially analytic). Here we find ourselves back in close contact with automaticians, this time with similar, synthetic aims.

† If this were the case, the system would have one or more stable states in addition to the two cyclic attractors.

Table II*a* tells us what is common to all acceptable structures, and it permits to identify individual solutions among the simplest ones.

For that, it is convenient to split the table into four subtables, one for each function; this gives Table II*b*. Using well-established methods (see, e.g., ref. 45), with a little practice one finds immediately simple combinations that fit the desiderata. The idea is to identify in the table one or

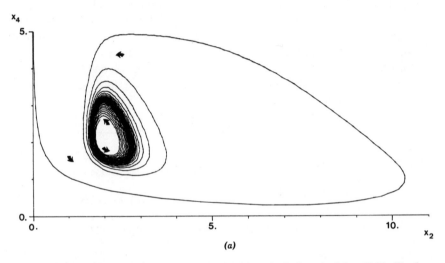

Fig. 1.   The "double cycle" system synthesized by a logical method (see Table II), then translated into a continuous description. Taking into account the fact that we took the k and $\theta = 1$, the equations used are

$$\dot{x}_1 = \frac{1}{1 + x_1^n} - k_{-1}x_1$$

$$\dot{x}_2 = \frac{1}{1 + x_1^n} \cdot \frac{1}{1 + x_4^n} + \frac{x_1^n}{1 + x_1^n} \cdot \frac{1}{1 + x_3^n} - k_{-2}x_2$$

$$\dot{x}_3 = \frac{1}{1 + x_1^n} \cdot \frac{1}{1 + x_2^n} + \frac{x_1^n}{1 + x_1^n} \cdot \frac{1}{1 + x_4^n} - k_{-3}x_3$$

$$\dot{x}_4 = \frac{1}{1 + x_1^n} \cdot \frac{1}{1 + x_3^n} + \frac{x_1^n}{1 + x_1^n} \cdot \frac{1}{1 + x_2^n} - k_{-4}x_4$$

The limit cycles were obtained first by a simple Euler interation on a pocket calculator, then confirmed by a Merson program of numeric integration, kindly performed by A. Goldbeter and O. Decroly. Other values of the parameters: $n = 3$; $k_{-i} = 0.3; 0.05; 0.1; 0.2$ (*a*) Trajectories toward one limit cycle "from outside" (initial state: $0.01 - 0 - 0 - 5$) and "from inside" (initial state: $0.01 - 2.1 - 1.1 - 2.3$).

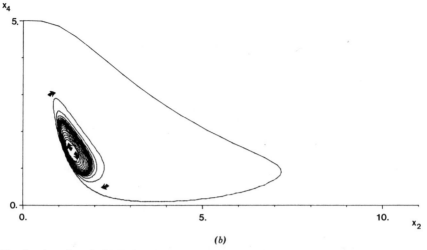

(b)

Fig. 1. (*continued*) (*b*) Trajectories toward the other limit cycle "from outside" (initial state: $5 - 0 - 0 - 5$) and "from inside" (initial state: $3.24 - 1.4 - 3.2 - 1.6$).

a small number of regions which together contain all the "1" and none of the "0", and can be described by a simple logic expression. For instance in the first part of Table II*b*, one sees that the boxed region contains all the "1," none of the "0," and corresponds to the region in which $\alpha = 1$; thus, $a = \alpha$ provides a simple solution as regards function a.

This procedure gives the following set of logical equations:

$$a = \alpha$$
$$b = \overline{\alpha}\overline{\delta} + \alpha\overline{\gamma}$$
$$c = \overline{\alpha}\overline{\beta} + \alpha\overline{\delta}$$
$$d = \overline{\alpha}\overline{\gamma} + \alpha\overline{\beta}$$

which must fit the requirements. The corresponding, complete state table is given in Table II*c*; it is easy to show that in this system, one falls into the first or the second cycle depending on whether the initial state has $\alpha = 0$ or $\alpha = 1$.

One may now ask: Can one find a *continuous* system of the same logical structure which, for proper values of the parameters, will go to either of two limit cycles running in opposite directions, depending on the initial state? One can derive from our set of logical equations a homologous set of differential equations as described in Section III.A. For the values of the parameters indicated in the legend, the system indeed goes to either of two limit cycles depending on the initial value of x_1 (see Fig. 1).

TABLE II. Synthesis of a Logical System Leading to the Two Cycles Given in the Text

(a) Partial state table

αβγδ	abcd
0000	$----$ (not 0000)
$00\overline{0}1$	0011
$001\overline{1}$	0010
$0\overline{0}10$	0110
$01\overline{1}0$	0100
0111	$----$ (not 0111)
$0\overline{1}01$	0001
$010\overline{0}$	0101
$11\overline{0}0$	1110
$110\overline{1}$	1100
1111	$----$ (not 1111)
$1\overline{1}10$	1010
$101\overline{0}$	1011
$10\overline{1}1$	1001
$1\overline{0}01$	1101
1000	$----$ (not 1000)

(b) The same, split into four tables, one for each function.

Function a:

γδ \ αβ	00	01	11	10
00	–	0	1	–
01	0	0	1	1
11	0	–	–	1
10	0	0	1	1

$a = \alpha$

Function b:

γδ \ αβ	00	01	11	10
00	–	1	1	–
01	0	0	1	1
11	0	–	–	0
10	1	1	0	0

$b = \overline{\alpha}\,\overline{\delta} + \overline{\alpha}\gamma$

Function c:

γδ \ αβ	00	01	11	10
00	–	0	1	–
01	1	0	0	0
11	1	–	–	0
10	1	0	1	1

$c = \overline{\alpha}\,\overline{\beta} + \alpha\overline{\delta}$

Function d:

γδ \ αβ	00	01	11	10
00	–	1	0	–
01	1	1	0	1
11	0	–	–	1
10	0	0	0	1

$d = \alpha\overline{\gamma} + \alpha\overline{\beta}$

(c) The complete state table

αβγδ	abcd
$0\overline{0}\,\overline{0}\,\overline{0}$	0111
$00\overline{0}1$	0011
$001\overline{1}$	0010
$0\overline{0}10$	0110
$01\overline{1}0$	0100
$0\overline{1}\,\overline{1}\,\overline{1}$	0000
$0\overline{1}01$	0001
$010\overline{0}$	0101
$11\overline{0}0$	1110
$110\overline{1}$	1100
$1\overline{1}\,\overline{1}\,\overline{1}$	1000
$1\overline{1}10$	1010
$101\overline{0}$	1011
$10\overline{1}1$	1001
$1\overline{0}01$	1101
$1\overline{0}\,\overline{0}\,\overline{0}$	1111

(a) Partial state table (incompletely specified functions).

(b) The same, split into four tables, one for each function. Boxes cover regions containing all the "1" and no "0" and which can be described by logical expressions as simple as possible.

(c) the complete state table corresponding to the simple solution identified under (b).

IV. DISCUSSION

In disciplines like biology, whose objects are very complex, there is a tendency to reason either in an entirely unformalized way, or, at the other extreme, in fully quantitative terms. Both attitudes have drawbacks as, on the one hand, it is very difficult to treat complex systems in the complete absence of formalism; on the other hand, a fully quantitative treatment of such systems often gives only an illusory impression of precision, as the values of the parameters had to be invented. We are interested in an intermediate attitude, which has been advocated, for example, by Thom[47] and by Glass and Kauffman[15] in which one is limited to the essential qualitative aspects of the dynamics of systems.

The approach I developed is a purely logical, fully asynchronous one, in which each element of the system is represented by a logical function associated with its rate of production (more generally, its *development*) and a logical variable associated with its concentration (more generally, its *actual state*). One reason for denoting them "function" and "variable," respectively, is that the systems are usually described by logical equations, each of which gives the value of the rate of production of an element as a function of the concentrations (internal variables) and of other ("input") variables. Much of the simplicity of the treatment comes from the fact that the value of the variable (α) associated with the concentration of an element of the system is related to the value of the corresponding function (a) (associated with the rate of production) by a simple time shift. When a and α are both 0 and a signal (an appropriate change in the value of a variable) makes $a = 1$, α will adopt the value 1 after a delay t_α unless a counterorder ($a = 0$ again) takes place before the order has been executed ($\alpha = 1$); but when this is the case, one merely ignores the order. Similarly, if a and α both equal 1 and a signal switches a from 1 to 0, α will follow after a delay $t_{\bar{\alpha}}$ (unless a counterorder . . .).

From the logical equations describing a system one can build a state table that gives the values of the functions for each combination of values of the variables defining the "state." And from the state table one can derive (by hand up to four to five variables, with a computer program if there are too many variables) the set of *logical trajectories* (sequences of logical states) conceivable for the system in question. For any such sequence one can proceed to an analysis of the conditions (initial state, values of the time delays) which will impose this sequence; in particular, one can proceed to a *logical stability analysis* of cyclic sequences. Final states can be *stable states* (defined by the equality of the vector ($\alpha_1, \alpha_2, \alpha_3, \ldots$) with the vector ($a_1, a_2, a_3, \ldots$) or *stable cycles* (perhaps also strange attractors). An essential point is that the logical identification of the number and nature of the attractors of a system can be performed "analytically," that is, without ascribing a numerical value to the time delays.

This logical analysis is self-sufficient. It is nevertheless often interesting to translate a set of logical equations into a set of differential equations of the same structure (in which logical functions have been replaced by sigmoids as proposed by Glass and Kauffman[15],) or vice versa, in order to establish a parallel between the two descriptions. The correspondence has been discussed in Sections I.B and III.A. In short, the Boolean "caricature" of a system is relatively easy to analyze. It provides a detailed qualitative description of the possible dynamical behaviors of a system, which, as far as I can tell, corresponds to the most "differentiated" set of behaviors of the corresponding continuous description; to be more specific, the qualitative behaviors uncovered by the logical analysis will be found in the continuous description for appropriate values of the parameters or delays, provided the interactions considered in the continuous description are sufficiently nonlinear. For instance, the logical analysis of a four-component system comprising a positive and a negative feedback loop tells that depending on the initial state and values of the delays the system has a choice between two distinct attractors, a stable state and a stable cycle; in addition it provides detailed indications about the shape of the stable cycle, including the prediction of multiple periodicities in appropriate conditions. After a somewhat discouraging purely continuous study, it has been found very convenient to discover the essential dynamical features of the system by logical analysis, then shift back to the continuous analysis.

Another attracting possibility has consisted of reversing the process: given an experimental (or desirable) behavior, and assuming that the relevant elements of the system have been identified, find interactions between these elements which will permit (or impose) exactly this behavior. Our method has been used to find networks of neurons which would behave exactly as found by Friesen and Stent[42] in the locomotion of the leech. In the present chapter, we also show that a relatively sophisticated desideratum (a four-dimensional system that would have a choice between two limit cycles running in opposite directions) can be fulfilled by a short logical analysis followed by translation from logical to differential equations; it has been very easy afterward to find values of the parameters such that the system goes to either of two limit cycles depending on its initial state.

A more difficult problem is to what extent this type of formalism will be applicable to chemical systems in which elements of the system transform into one another rather than merely affecting the rates of synthesis of each other. Interesting suggestions by my colleagues Jeener and Nicolis (see Section III.D) might greatly simplify the situation. But, as already remarked, this type of application will at best still require much more work. Finally, one might mention that the logical method described here has been applied already to systems involving feedback loops in various disciplines other than genetics and even than biology: see, or instance, the essays of Boon and de Palma[48] in urbanism, of de Palma, Stengers, and Pahaut[49] in operational research and of C. Nicolis[50] in meteorology. It is comforting to note that in these different cases the formalism originally developed for genetic systems could be used as such.

References

1. R. Thomas, *J. Theor. Biol.*, **42**, 563–585 (1973).
2. R. Thomas and P. Van Ham, *Biochimie*, **56**, 1529–1547 (1974).
3. R. Thomas, *Lecture Notes in Biomathematics*, **49**, 189–201 (1983).

4. A. Novick and M. Wiener, *Proc. Natl. Acad. Sci. USA*, **43**, 553–566 (1957).
5. M. Cohn and K. Horibata, *J. Bact.*, **78**, 601–623 (1959).
6. H. Eisen et al., *Proc. Natl. Acad. Sci. USA.*, **66**, 855–862 (1970).
7. Z. Neubauer and E. Calef, *J. Mol. Biol.*, **51**, 1–13 (1970).
8. T. M. Sonneborn, (1975) in *Handbook of Genetics*, R. C. King, ed., Vol. 2, pp. 469–594.
9. N. Glansdorff and I. Prigogine, *Thermodynamics of Structure, Stability and Fluctuation*, Wiley-Interscience, New York, 1971.
10. G. Nicolis and I. Prigogine, *Self-Organization in Nonequilibrium Systems*, Wiley-Interscience, New York, 1977.
11. J. Richelle, *Bull. Cl. Sci. Acad. Roy. Belg.*, **66**, 890–912 (1980).
12. L. Glass and S. A. Kauffman, *J. Theor. Biol.*, **34**, 219–237 (1972).
13. R. Thomas, A. M. Gathoye, and L. Lambert, *Eur. J. Biochem.*, **71**, 211–227 (1976).
14. R. Thomas, *J. Theor. Biol.*, **73**, 631–656 (1978).
15. L. Glass and S. A. Kauffman, *J. Theor. Biol.*, **39**, 103–129 (1973).
16. P. Van Ham and I. Lasters, *J. Theor. Biol.*, **72**, 269–281 (1978).
17. N. Rashevsky, *Mathematical Biophysics*, University of Chicago Press, 1948.
18. M. Sugita, *J. Theor. Biol.*, **4**, 179–192 (1963).
19. U. Kling and G. Szekely, *Kybernetik*, **5**, 89–103 (1968).
20. S. A. Kauffman, *J. Theor. Biol.*, **22**, 437–467 (1969).
21. S. A. Kauffman, *J. Theor. Biol.*, **44**, 167–190 (1974).
22. L. Glass, *J. Chem. Phys.*, **63**, 1325–1335 (1975).
23. R. N. Tchuraev and V. A. Ratner, in *Studies on Mathematical Genetics*, V. A. Ratner, ed., Inst. Cytol. Genet. Press, Novosibirsk (in Russian), p. 5, 1975.
24. L. Glass and J. S. Pasternak, *Bull. Math. Biol.*, **40**, 27–44 (1978).
25. L. Glass, *J. Theor. Biol.*, **54**, 85–107 (1975).
26. D. A. Huffman, *J. Franklin Inst.*, **257**, 161–189 (1954).
27. J. Florine, *La Synthèse des Machines Logiques et Son Automatisation*, Dunod, Paris, 1964.
28. P. Van Ham, Thesis, University of Brussels, 1975.
29. M. Milgram, Thesis, University of Compiègne, France, 1982.
30. R. Thomas, ed., "Kinetic logic: A Boolean approach to the analysis of complex regulatory systems," *Lecture Notes in Biomathematics*, Vol. 29, Springer-Verlag, Berlin, 507 pp.
31. J. Leclercq and R. Thomas, *Bull. Cl. Sci. Acad. Roy. Belg.*, **67**, 190–225 (1981).
32. P. Van Ham and J. L. De Houck, *Lecture Notes in Biomathematics*, **29**, 149–163 (1979).
33. B. C. Goodwin, *Adv. Enzyme Regul.*, **3**, 425–438 (1965).
34. J. S. Griffith, *J. Theor. Biol.* **20**, 202–216 (1968).
35. C. Walter, *Quantitative Biology of Metabolism*, A. Locker, ed., Springer-Verlag, Berlin, 1967, pp. 38–44.
36. P. Van Ham, *Lecture Notes in Biomathematics*, **29**, 142–148 (1979).
37. J. Richelle, *Lecture Notes in Biomathematics*, **29**, 281–325 (1979).
38. R. Thomas, G. Nicolis, J. Richelle, and P. Van Ham, *Lecture Notes in Biomathematics*, **29**, 345–352 (1979).

39. F. Robert, Rapport de recherche no. 163, IMAG, Grenoble, 1979.

40. E. Goles- Chacc, Rapport de recherche no. 157, IMAG, Grenoble, 1979.

41. R. Thomas, *Springer Series in Synergetics,* **9,** 180–193 (1981).

42. W. D. Friesen and G. S. Stent, *Biol. Cybernetics,* **28,** 27–40 (1977).

43. R. Thomas, *Biologie Moléculaire: Progrès Récents et Perspectives,* Editions de l'Université Libre de Bruxelles, 1983.

44. P. Van Ham, *Bull. Cl. Sci. Acad. Roy Belg.,* **68,** 267–294 (1982).

45. A. Leussler and P. Van Ham, *Lecture Notes in Biomathematics,* **29,** 62–106 (1979).

46. J. Richelle, *Bull. Cl. Sci. Acad. Roy. Belg.,* **63,** 534–546 (1977).

47. R. Thom, *Modèles Mathématiques de la Morphogénèse,* C. Bourgeois, Paris, 1980.

48. F. Boon and A. de Palma, *Lecture Notes in Biomathematics,* **29,** 402–439 (1979).

49. A. de Palma, I. Stengers, and S. Pahaut, R.A.I.R.O. *Rech. Oper.,* **16,** 155–167 (1982).

50. C. Nicolis, *Quart. J. R. Met. Soc.,* **108,** 707–715 (1982).

INDEX